Graphing Calculator Lab Manual

For the Allyn J. Washington Series in

Basic Technical Mathematics

Sixth Edition

Robert E. Seaver

Lorain County Community College

Addison-Wesley Publishing Company

Reading, Massachusetts • Menlo Park, California • New York
Don Mills, Ontario • Wokingham, England • Amsterdam • Bonn
Sydney • Singapore • Tokyo • Madrid • San Juan • Milan • Paris

I dedicate this book to
Jed, Cody, Ethan, and Araenae,
and other grandchildren yet to come.

Reproduced by Addison-Wesley from camera-ready copy supplied by the author.

ISBN 0-201-63129-6
1 2 3 4 5 6 7 8 9 10- VG-98979695

Preface

In recent years, the development of the graphing calculator has made a major impact on the teaching of mathematics. The graphing calculator enables students to better understand mathematical concepts, visualize functions, perform matrix operations, and perform statistical calculations. Such a tool is particularly well suited to the engineering technologies.

The main purpose of this work is to provide the information necessary to enhance the teaching of technical mathematics through use of the graphing calculator. A secondary purpose is to provide a background that will enable the student to efficiently use the graphing calculator in other classes and in future employment. Most of this material has been successfully used for several years in technical mathematics classes at Lorain County Community College.

In this material, use of the graphing calculator is developed through the learning of procedures. The procedures are presented in an order that corresponds to their use in a typical technical mathematics course. Experience has shown that students acquire these procedures quite quickly and rarely need to refer back to procedures previously learned. However, if a reference is needed, a summary of calculator procedures is available in the appendix.

This manual is written at a level that should be understood by a technical mathematics student with a good basic knowledge of algebra. Each section contains a brief explanation of the material under consideration, example problems, and related exercises. The explanations are not intended to be a complete, in-depth coverage of topics, but only to serve as background material for the calculator procedures and exercises.

The material has been keyed to corresponding material in the Allyn J. Washington's texts: <u>Technical Mathematics</u> and <u>Technical Mathematics with Calculus</u>, Sixth Edition. Sections in the text that correspond to material in this manual are indicated under section headings by the word *Washington* followed by a section number. For example: *(Washington, Section 3.4)*

In addition, certain examples and exercises in this manual correspond to examples and exercises in Washington's text. When such a correspondence exists, the example or exercise is indicated by the symbol ♦ in the left margin. The corresponding text example is indicated by the word *Washington* followed by the section in which the example appears and the example number. For example: *(Washington, Section 3.5, Example 3)* A corresponding text exercise is indicated by the word *Washington* followed by the exercise set and the problem number. For example: *(Washington, Exercise 3.4, #37)* These examples and exercises are similar to the corresponding examples and exercises in the text. However, in many cases, changes were made in the problems to better incorporate the use of the graphing calculator.

Explanations, examples, and exercises are generally calculator independent. The calculator procedures are directed toward the Texas Instruments Graphing Calculators TI-82 and TI-85. Appendices provide a summary of procedures for the TI-81, TI-82, TI-85, and Casio 7700 graphing calculators. Students using other makes of calculators will need to refer to their user's manual.

It has been a pleasure writing this material as well as using it with mathematics classes. It is my desire that it be error free. However, being realistic, errors certainly will surface. I apologize in advance for any errors and my hope is that they will cause only minor inconvenience.

I appreciate the help and encouragement given me by Bill Poole, Jason Jordan, and others at Addison-Wesley. I also wish to acknowledge the love, patience, encouragement, and understanding given by my wife Janet.

RES

Contents

Chapter 1

Introduction to a Graphing Calculator

1.1 Differences from a Standard Scientific Calculator.

Modern technology provides tools that were never available at any time in the past. One such tool is the graphing calculator. The graphing calculator, in addition to doing all the calculations normally done on a scientific calculator, readily graphs mathematical functions, does statistical calculations, and performs matrix operations. It is an extremely valuable tool and is, in fact, a miniature computer. It can also be programmed to solve mathematical problems in much the same way that a computer can be programmed.

There are several differences between a graphing calculator and a standard scientific calculator. An obvious difference in the larger screen that not only permits the viewing of a graph but also allows the user to see a complete problem and not just the answer. As we shall see, this allows us to readily make changes in the expression we have entered if an error is detected. Another difference is the keyboard which contains additional keys not normally found on a standard scientific calculator.

The material in this supplement applies specifically to the Texas Instruments Graphing Calculators models TI-82 and TI-85. Most of the discussion may also be applied to other graphing calculators. However, since with other calculators certain operations may be handled differently, the user is referred to the manual that came with the calculator.

For information regarding turning the calculator on and off, clearing all memory, adjusting the contrast of the screen, and clearing the screen, the user is referred to Basic Procedures B1, B2, B3, and B4 in the appendices of this supplement.

1.2 On Working with the Calculator

The screen on the graphing calculator is made up of dots called **pixels**. Characters or graphs appear on the screen when certain of these pixels are turned on. A typical graphing calculator screen contains 96 pixels in the horizontal direction and 64 pixels in the vertical direction. When the graphing calculator is turned on and the screen is clear, there should be a blinking rectangle or line on the screen. This is referred to as a

cursor and indicates the location where a letter or number will appear when entered from the keyboard. On the graphics screen, the cursor may appear as a cross. On the upper right of the keyboard are four arrow keys called the **cursor keys.** These keys are used for moving the cursor around the screen.

Many of the keys on the keyboard, in addition to their primary use, are also used to display alpha characters or to perform secondary actions. **Characters** are any numbers, letters, or symbols that may be displayed on the calculator screen. The **alpha characters** are letters and symbols that may be used as variables or in writing programs. Characters are normally displayed in up to eight rows with up to 16 characters across the screen. **Secondary operations** are additional functions performed by many of the keys. The alpha characters appear above and to the right of a key and the secondary operations appear above and to the left of a key. The methods of accessing these alpha characters or secondary operations are given in Procedure C1.

Procedure C1. To perform secondary operations or obtain alpha characters.

1. To perform secondary operations:
   a) Press the key: 2nd
   b) Then, press the key for the desired action
      (For example: To find the square root of a number, first press the key: 2nd followed by the key: x^2. Then, enter the number and press ENTER.)
2. To obtain alpha characters:
   a) Press the key: ALPHA
   b) Then, press the key for the desired character.
      (For example: To display the letter A on the screen, press the key: ALPHA then press the key MATH or the TI-82 or the key LOG on the TI-85)

   Note: The alpha key may be locked down on the TI-82 by first pressing the key 2nd then pressing the key: ALPHA. On the TI-85, press the key ALPHA twice.

Graphing calculators are menu oriented and when certain keys are pressed, menus appear on the screen. A **menu** is a list of possible options from which we may make a selection in much the same way as we would select an item from a dinner menu in a restaurant.

On occasion, a word, item, number, or area of the screen is highlighted. A **highlighted area** is that area with the background shaded.

Procedure C2. **Selection of menu items**

A menu on the TI-82 calculator appears as a list of numbers, each followed by a colon and a menu item. A menu item may be selected by:

1. Pressing the number of the item desired

or

2. Moving the highlight to the item desired by pressing the up and down cursor keys, then pressing the key ENTER.

(Example: To find the cube root of 3.375, press the key MATH, leave MATH highlighted in the top row and select 4:$\sqrt[3]{}$. Then enter the number 3.375 and press ENTER. The result is 1.5.)

In some cases more than one menu may be accessed by pressing the same key. This is indicated by more than one title appearing at the top of the screen with one of the words highlighted. As the right and left arrow keys are used to move the highlight from one title to another. different menus appear on the screen. Thus, first select the correct menu, then the desired item.

(Example: To find the largest integer which is less than or equal to the number 2.15, first press the key MATH, move the highlight in the top row to NUM, select 4:Int, enter the number 2.15, and press ENTER. The answer is 2.)

To exit from a menu without selecting an option, press QUIT.

A menu on the TI-85 calculator appears as a list on the bottom of the screen. A selection is made by pressing the function key just below the item desired. The function keys are the keys labeled F1, F2, F3, F4, and F5.

In some cases, election of a menu item will produce a secondary menu. Items are selected from the secondary menu by pressing the appropriate function key.
If the symbol → appears on the right of a menu, there are additional items in the menu. To see these additional items, press the key: MORE
To exit from a menu to the previous menu or to the calculations screen, press the key: EXIT

(Example: To find the largest integer less than or equal to 2.095, press the key: MATH, then press the key F1 for NUM, and from the secondary menu, select **int** by pressing the key F4. Then enter the number 2.095 and press ENTER. The result is 2.)

1.3 Numerical Calculations and Error Correction

We are now ready to use the calculator to do numerical calculations in much the same way as may be done with any scientific calculator. But first a few comments about the order of arithmetic operations.

If we enter the arithmetic expression $3^4 + 5 * 2 - 5 \div 3$ into a scientific calculator (or into a computer) there is a specific order in which the calculator will perform the operations of addition, subtraction, multiplication, division and of finding exponents and roots. This order should also be followed when doing a calculation by hand.

Order of operations:

First: Find all powers (exponents) and roots.
Second: Perform all multiplication's and divisions in order from left to right.
Third: Perform all additions and subtraction's in order from left to right.

If parenthesis appear in the expression, the operations inside the parenthesis are performed first. If more than one set of parenthesis appears, the expression in the innermost parenthesis is performed first, then the second innermost parenthesis, etc. Within a set of parenthesis, the given order of operations is followed.

Example 1.1: Evaluate by hand $3^4 + 5 \times 2 - 6 \div 3$.

Solution: 1. First do the power to obtain

$$81 + 5 \times 2 - 6 \div 3$$

2. Second do the multiplication's and divisions

$$81 + 10 - 2$$

3. Finally do additions and subtraction's

$$89$$

On the calculator, if more than one exponent operation appears in order, the operations are performed from left to right. (In a computer program, exponent operations are normally performed from right to left.) Thus, if you enter 2^2^3 on your calculator (the symbol ^ is used to find exponents on the calculator), and press ENTER, you will obtain 64 since 2^2^3 = 4^3 = 64. It is best when using more than one exponent operation in a row to use parenthesis to indicate which operation is to be done first.

Next we perform the calculation of Example 1.1 on a calculator. When entering an expression into the calculator we use the arithmetic operation keys on the right hand side. Procedure C3 indicates how to perform arithmetic operations.

Procedure C3. To perform arithmetic operations.

1. To add, subtract, multiply or divide press the appropriate key: $+, -, \times,$ or $\div$. On the graphing calculator screen, the multiplication symbol will appear as * and the division symbol will appear as /. If parenthesis are used to denote multiplication it is not necessary to use the multiplication symbol. (The calculation 2(3 + 5) is the same as 2×(3 + 5).)
2. To find exponents use the key ^
   (Example: 2^4 is entered as 2^4)
   For the special case of square we may use the key x^2.
3. After entering the expression press the key ENTER to perform the calculation.

Notes: a. On the graphing calculator, there is a difference between a negative number and the operation of subtraction. The key (-) is used for negative numbers and the key – is used for subtraction.

b. Enter numbers in scientific notation by using the key EE.
(Example: To enter 2.5×10^5, enter 2.15, press EE, enter 5.)

A graphing calculator has the advantage that the complete entered expression is visible on the screen. This allows us to check the expression to see if it has been entered correctly or to quickly and easily change a value in the expression.

Procedure C4. To correct an error or change a character within an expression.

1. (a) To make changes in the current expression:
   Use the left cursor key to move the cursor to the position of the character to be changed.
   (b) To make changes in the last expression entered, after the ENTER key has been pressed, obtain a copy of the expression by pressing the key ENTRY. Then, use the cursor keys to move the cursor to the character to be changed.
2. Characters may now be replaced, inserted, or deleted at the position of the cursor.
   (a) To replace a character at the cursor position, just press the new character.

(b) To insert a character or characters in the position the cursor occupies,
Press the key: INS.
Then, press keys for the desired character(s).

(c) To delete a character in the position the cursor occupies, press the key: DEL

Example 1.2: Use the calculator to evaluate $3^4 + 5 \times 2 - 6 \div 3$ then, evaluate the expression $3^5 + 5 \times 2 - 6 \times 3$.

Solution:

1. Enter the first expression into the calculator exactly as given and perform the calculation (see Procedure C3). The answer of 89 should appear one line down and on the right of the screen.
2. The second expression differs from the first only in the exponent of the number 3. We need to change the exponent from a 4 to a 5. Use Procedure C4 to make changes in the last expression. Move the cursor to the 4 and replace with a 5. Press ENTER to perform the calculation. The value 251 for the second expression of 251 should appear on the right of the screen.

In Example 1.2, we made use of Procedure C4 to replace a character in an expression. This same procedure may be used to correct an error within an expression that we have just entered.

We need to take special care when working with fractions. Consider the evaluation of the quantity $\frac{3+9}{2}$. If we enter this as $3 + 9 \div 2$, we obtain the value is 7.5 since the calculator is following the order of operations and first divides 9 by 2.. In order to have the 3 added to the 9 before division, it is necessary to insert parenthesis and enter the expression as $(3+9) \div 2$. This gives the correct answer of 6. In a fraction, expressions in the numerator or in the denominator must be evaluated first before doing division. For purposes of entering fractions into a calculator a numerator or denominator containing more than one quantity should be thought of as being in parenthesis. Example 1.3 illustrates the use of parenthesis and of changing numerical values within an expression.

Example 1.3: Evaluate the expressions $6500\left(1+\frac{0.06}{2}\right)^8$, $6500\left(1+\frac{0.06}{4}\right)^{16}$, and $6500\left(1+\frac{0.06}{12}\right)^{48}$. Round answers to the nearest hundredth. These expressions calculate the amount of money in a savings account after 4 years for an initial deposit of $6500, earning interest at the rate of 6% per year and compounded 2 times a year, 4 times a year, and 12 times a year respectively.

Solution:

1. Use Procedure C3 to enter the first expression $6500\left(1+\frac{0.06}{2}\right)^{8}$ into the calculator and perform the calculation. The result should be: 8234.005529 which we round this to 8234.01.
2. To evaluate the second expression we need only change two of the numbers. Use Procedure C4 to change the 2 to a 4 and the 8 to 16 and perform the second calculation. The answer is 8248.40606 and we round this to 8248.41.
3. To evaluate the third expression, we now need to change the 4 to a 12 and the 16 to a 48. Again use Procedure C4 to make these changes and perform the calculation. The answer is 8258.179547 or 8258.18.

Special mathematical functions may require use of special menus. We will consider only those functions which are important to us at this time.

Procedure C5. Selection of special mathematical functions.

1. To raise a number to a power:
   Enter the number
   Press the key ^
   Enter the power
   Press ENTER.
2. To find the root of a number (other than square root):

   On the TI-82:
   Enter the root index
   Press the key MATH
   With Math highlighted,
   Select $\sqrt[x]{\ }$
   Enter the number
   Press ENTER

   On the TI-85:
   Enter the root index
   Press the key MATH
   Select MISC
   Press MORE,
   From the secondary menu,
   Select $\sqrt[x]{\ }$
   Enter the number
   Press ENTER.
3. To find the factorial of a positive integer:

   On the TI-82:
   Enter the integer
   Press the key MATH
   Highlight PRB
   Select 4:!
   Press ENTER

   On the TI-85:
   Enter the integer
   Press the key MATH
   Select PROB
   From the secondary menu,
   select !
   Press ENTER.
4. To find the largest integer less than or equal to a given value (This is called the greatest integer function and is indicated by square brackets [ ]):

On the TI-82:
Press the key MATH
Highlight NUM
Select 4:Int
Enter the value
Press ENTER

On the TI-85:
Press the key MATH Select NUM
From the secondary menu, select int
Enter the value
Press ENTER.

5. For the number π, press the key for π on the keyboard.
6. To find the absolute value of a given value:

On the TI-82:
Press the key ABS
on the keyboard.

On the TI-85:
Press the key MATH
From the menu, select NUM
Select int
Enter the value and press ENTER.

Example 1.4: Find the square, square root, cube, and cube root of π. Also, find 8! and [-1.78]. (the greatest integer function.) Round answer to four decimal places.

Solution:

1. Find the square of π by:

   Entering π, pressing the key x^2, and pressing ENTER.
   The answer is 9.8696.
2. Find the square root of π by

   Pressing the key $\sqrt{\ }$, entering π, and pressing ENTER. The answer is 1.7725.
3. Find the cube of π by using Procedure C5 to raise a number to a power. The answer is 31.0063.
4. Find the cube root of π by using Procedure C5 to find the root of a number. The answer is 1.4646.
5. Find the factorial of the integer 8 by using Procedure C5. The answer is 40320.
6. Find the greatest integer function of −1.78 by using Procedure C5. The answer is -2.

1.4 About Significant Digits, Precision, and Accuracy.

In approximate numbers, not all the digits may have meaning. If a technician were to measure the resistance of a resistor with an ohmmeter that was only accurate to a tenth of a ohm, it would be pointless to write down that the resistance is 60.125 ohms. Instead, the technician should write down that the resistance 60.1 ohms. The number is rounded to indicate the accuracy of the measurement. The digits 2 and 5 are meaningless in this case.

When working with approximate numbers it is assumed that only the digits that have meaning appear. Such digits are referred to as **significant digits**. We will assume that any not zero digit is a significant digit and the digit zero is significant if

(1) if it appears between two other significant digits

or

(2) if it is to the right of a significant digit and is to the right of the decimal point.

Example 1.5: Give the number of significant digits in each of the numbers (a) 1203., (b) 2.3050, (c) 0.00200, and (d) 124000.

Solution:

(a) The number 1203 has 4 significant digits. The zero is significant since is it between two other significant digits.
(b) The number 2.3050 has 5 significant digits. The first zero appears between two significant digits and the second zero appears to the right of a significant digit and to the right of the decimal point.
(c) The number 0.00200 has 3 significant digits. The zeros in front of the 2 are not significant and the two zeros behind the 2 are significant since they are to the right of a significant digit and to the right of a decimal point.
(d) The number 124000 has 3 significant digits. In this case the three zeros are not significant even though they are to the right of a significant digit, since they are to the left of the decimal point.

The words accuracy and precision are closely tied to the idea of significant digits. **Accuracy** has to do with the number of significant digits in a value and **precision** has to do with the location of the rightmost significant digit. The larger the number of significant digits, the more accurate the number. The more to the right, in relation to the decimal point, the rightmost significant digit appears, the more precise the number.

Example 1.6: Consider the numbers 123., 0.00020, and 2034.2.

(a) Which of the numbers is the most accurate and which is least accurate?
(b) Which is most precise and which is least precise?

Solution:

(a) The number 2034.2 is most accurate since it contains the most significant digits (five). The number 0.00020 is least accurate since it contains the fewest number of significant digits (two).

(b) The number 0.00020 is most precise since the rightmost significant digit is farthest to the right of the decimal point. The number 123. is least precise since the rightmost significant digit is farthest to the left relative to the decimal point.

When doing calculations the results often contain a large number of digits, many of which are not really significant. This is particularly true when performing calculations on a calculator where results may contain 8 or more digits. In order to obtain answers that have the proper degree of accuracy or precision, we make use of certain guidelines.

Guidelines for Rounding:

1. When adding or subtracting approximate numbers, the result should not be more precise than the least precise of the numbers being added or subtracted.
2. When multiplying or dividing approximate numbers, the result should not be more accurate than the least accurate of the numbers being multiplied or divided.
3. When finding roots or powers, the result should not be more accurate than the numbers involved.
4. The following table gives comparative accuracy for working with trigonometric functions.

Angle (in degrees) to nearest	Angle (in radians) to	Value of trigonometric functions to
degree	2 s.d.	2 s.d .
0.1 degree	3 s.d.	3 s.d.
0.01 degree	4 s.d.	4 s.d.
0.001 degree	5 s.d.	5 s.d.

Of course, not all numbers are approximate. For example, when calculations involve the numbers of quarts in a gallon or the number of meters in a kilometer, the numbers involved (4 or 1000) are by definition **exact**. Other exact numbers may occur in formulas such as in $F = \frac{9}{5}C - 32$. When performing calculations involving both exact and approximate numbers, the rules above are used only in reference to the approximate numbers. Exact numbers are assumed to be "infinitely accurate and precise".

1.5 A Word About Calculator Mode

The **mode** of a calculator provides internal settings for the calculator and gives criteria for doing calculations and graphing of functions.

Procedure C6. Changing the mode of your calculator.

Press the key: MODE
You will see several lines. The items which are highlighted in each line set a particular part of the mode.

The first two lines set the way that numbers are displayed on the screen. In the first line, Norm is used for normal mode, Sci for scientific notation, and Eng for engineering notation. In the second line, Float causes numbers to be displayed in floating point form with variable number of decimal places and if one of the digits 0 to 9 is highlighted, all calculated numbers will be displayed with that number of decimal places.

The third line is used for trigonometric calculations and indicates if the angular measurements are in degrees or radians. Other lines on this screen will be considered later.

To change mode:

1. Use the cursor keys to move the cursor to the item desired.
2. Press ENTER. The new item should now be highlighted.
3. To return to the calculations screen, press the CLEAR or QUIT key.

(Example: To change the mode so that numbers are displayed in scientific notation with two decimal places, highlight Sci in the first line and press ENTER, then highlight 2 in the second line and press ENTER. Then, return to the calculations screen.)

1.6 Working with Sets of Numbers

On some calculators it is possible to work with sets of numbers called **lists**. Lists provide a convenient way to handle a set of numbers that are related.

Procedure C7. Storing a set of numbers as a list.

On the TI-82:
Up to six lists (of up to 99 values each) may be stored using the names $L_1, L_2, L_3, L_4, L_5,$ and L_6

Method I:
1. Press the key {
2. Enter the numbers to be in the list separated by commas.
3. Press the key }
4. Press the key STO▷
5. Press the key for the list name and press ENTER.

Method II:
1. Press the key STAT
2. From the menu, select 1:EDIT
3. Move the highlight to the list and position where numbers are to be entered, and enter the number. The entered value will appear at the bottom of the screen. Press ENTER to place the number in the list.
4. When all values in the list have been entered, press QUIT to exit from the list table.

Note: A value may be changed by moving the highlight in the table to that value, entering the correct value, and pressing ENTER.

On the TI-85:
Lists (of any length) may be stored using a list name of up to eight characters. The first character of the name a must be a letter.

Method I:
1. Press the key LIST.
2. Select {
3. Enter the numbers to be in the list separated by commas.
4. Select }
5. Press the key STO▷
6. Enter the list name and press ENTER.

METHOD II:
1. Press the key LIST
2. Select EDIT
3. To enter a new list, enter the new name after Name= . To change values in a stored list enter the name of the list or select from the list of names.
4. Enter the values as elements of the list. After e1= enter the first value and press ENTER, after e2= enter the second value and press ENTER, etc. When finished, press QUIT.

Note: A value may be changed by moving the cursor to that value and entering the new value.

Procedure C8. Viewing a set of numbers.

On the TI-82:
Press the key for the name of the list and press ENTER.
or

On the TI-85:
Enter the name of list and press ENTER.
or

1. Press the key STAT
2. Select 1:EDIT
3. Move the highlight to the list.
4. Press QUIT to exit.

1. Press the key LIST
2. Select NAMES, then select the list name and press ENTER or
   Select EDIT, enter or select the list name, and press ENTER.

The arithmetic operations of addition subtraction, multiplication, or division may be performed on lists. When operations are performed on lists the actions are performed on the corresponding elements of the lists. For example, the product {1, 2, 3, 4}{5, 6, 7, 8} will yield {5, 12, 21, 32}. It is also possible to operate on all elements of a list by a constant. For example, the sum {1, 2, 3, 4} + 5 adds 5 to each element of the list and gives {6, 7, 8, 9}and the square root of {1, 2, 3, 4} will yield {1, 1.41, 1.73, 2} (rounded to three significant figures).

Procedure C9. To perform operations on lists.

1. To add, subtract, multiply or divide lists, enter the list or the name of a list, press the appropriate key (+, −, ×, or ÷), enter the second list or the list name and press ENTER. Lists must be of the same length in order for these operations to be performed.
2. To perform an operation on all the elements of a list, enter a value, press the appropriate key (+, −, ×, or ÷), enter the list or list name and press ENTER.
3. Functions such as finding the squares or square roots, may be performed on each element of a list by performing the function of the list or list name.

Note: For multiplication it is not necessary to use the symbol ×.

Example 1.7: Find the sum of the set of numbers 12.5, 16.7, 22.1, and 16.3 and the set of numbers 32.1, 22.8, 45.6, 23.1 and then find 11.5% of each number in the result by multiplying the result by 0.115.

Solution:

1. Enter the set of number 12.5, 16.7, 22.1, and 16.3 in one list using the list name L1 and the set of numbers 32.1, 22.8, 45.6, and 23.1 into a second list using the list name L2 as in Procedure C7.
2. Find the sum of the two lists by adding the two list names, L1 + L2, and pressing ENTER. This sum is stored under Ans. The result is the list 44.6, 39.5, 67.7, and 39.4.
3. To find 11.5% of the product, multiply Ans by 0.115 (this may be done by first pressing × and then entering .115) and press ENTER. The result, rounded to three decimal places, is the list 5.13, 4.54, 7.79, and 4.53.

Exercise 1.1

Use the graphing calculator to evaluate each of the following. Round answers to the proper number of significant digits. Consider integers as exact numbers.

1. $3.1526 + 233.4 + 122.156$
2. $25.2 + 6.3014 + 231.34 + 0.066$
3. $23.125 - 7.24$
4. $9.1234 - 0.852$
5. $-13.271 + 6.344 - 6.25$
6. $-212.3 + 156.57 + 193.682$
7. $(0.034)(20.2)$
8. $(52.37)(6.1359)$
9. $607 \div 28$
10. $-10635 \div 7.28$
11. $\dfrac{23.2 + 16.7}{18.9 - 6.25}$
12. $\dfrac{-8.195 + 23.42}{6.576 + 19.1732}$
13. $\sqrt{23.4}$
14. $\sqrt{30.0^2 + 40.0^2}$
15. $\sqrt{12.2^2 - 3.75^2}$
16. $16.7\sqrt{245.7}$
17. $(3.7)(0.125) + 6.783$
18. $(8.72 \div 3.19) + (0.03527)(123.4)$
19. Find the average of 74.2, 63.5, and 91.2. You discover the number 63.5 should have been 68.5. Make the change and determine the new average.
20. Find the average of 83, 72, 91, 95, 67, 81, and 98. You discover the number 67 should have been 76. Make the change and determine the new average.
21. $6.2^2 + (3.5)(4.221)$
22. $3.5^2 + (62)(3.75)(83.2 + 16.4)$
23. $6.514^3 + \sqrt{82.15}$
24. $8.74^4 - \sqrt{17.26}$
25. If $I = Prt$, where $P = 5225$ and $t = 0.750$, then find I for the following values of r: 0.050, 0.055, 0.060, and 0.065
26. If $F = \frac{9}{5}C + 32$, then find F for the following values of C: 10.2, 15.0, 20.0, 25.5, and 31.2.
27. Modify the formula in exercise 26 and find C for the following values of F: 32.0, 68.0, 98.6, 100.0, and 155.5.

28. If $R = \dfrac{R_1 R_2}{R_1 + R_2}$, then find R if
    (a) if $R_1 = 16.3$ and $R_2 = 152.5$.
    (b) if $R_1 = 155.6$ and $R_2 = 226.3$.
29. $3.45^3 \pi$
30. $\dfrac{0.1837}{7.34^3}$
31. $\sqrt[3]{625 - 24.235^2}$
32. $\sqrt[3]{2255 + 45.27^2}$
33. $63.7\sqrt[4]{82.4}$
34. $0.5123\sqrt[5]{62.4 + 25.3^3}$
35. $(1.772)\,(3.1563^4)$
36. $\dfrac{4(62.34^2)}{3(1.25^4)}$
37. 12!
38. 3!4!
39. $\dfrac{8!}{7!2!}$
40. $\dfrac{14!}{5!9!}$
41. The number of ways in which n digits can be arranged in an order (with no digits repeating) is n!. How many ways can 7 digits be arranged?
42. In how many way can the 26 letters of the alphabet be arranged (see previous exercise)?
43. Evaluate $\dfrac{(4 \times 62)}{(3 \times 125)}$. Are both sets of parentheses needed?
44. Evaluate $\dfrac{(23.0 \div 3.12)}{(3.5 \div 12.7)}$. Are parentheses needed?
45. Find [2.999]
46. Find [−1.999]
47. Find [3.5/0.8]
48. Find [−12/9]

In Exercises 49-56, use lists.

49. Store {34, 45, 62, 23} as one list and {76, 23, 43, 45} as a second list. Find the sum, difference, and product of the two lists. Also, divide the first list by the second.
50. Store {22.4, 34.1, 55.6, 35.4, 22.1} as one list and {44.3, 22..7, 87.6, 56.7, 89.0} as a second list. Find the sum, difference, and the product of the two lists. Also, divide the first list by the second.
51. Store {12, 15, 42, 33, 68} as a list. Multiply the list by π, divide the list by 6.5, and subtract 17 from the list.
52. Store {22.5, 34.6, 89.7, 78.3} as a list. Multiply the list by 22.4, divide the list by 2.25, and add 23.5 to the list.
53. Use lists to find the square and square root of each of the numbers 22.5, 11.6, 32.5, and 64.0.
54. Use lists to find the cube and cube root of each of the numbers 22.5, 11.6, 32.5, and 64.0.
55. Store {20.0, 0.00, –40.0, 37.0, 100.0} as a list. Multiply the list by (9/5) and add 32 to the result. What to these lists represent?
56. A clerk in the store room determines that the number of washers 0.50 inches in diameter is 135 and the number of washers 0.25 inches in diameter is 238, the number of washers 0.125 inches in diameter is 25, and the number of washers 0.375 inches in diameter is 78. The value of each of the 0.50 inch washers is $2.10, the 0.25 inch washers is $1.75, the 0.125 inch washers is $0.95 and the 0.375 inch washers is $2.55. Set up two lists and multiply them to determine the value of the washer of each size in stock.

The slope of a straight line is given by $m = \dfrac{y_2 - y_1}{x_2 - x_1}$ where (x_1, y_1) is one point on a line and (x_2, y_2) is a second point on the line.

57. Find the slope of the line through the points (7.2, –3.1) and (3.11, –5.0).
58. Find the slope of the line through the points (–1.63, 2.50) and (3.72, –1.233).

The distance between two points (x_1, y_1) and (x_2, y_2) is given by the formula $\sqrt{(x_2 - x_1)^2 + (y_2 - y_1)^2}$.

59. Find the distance between the points (2.334, 1.77) and (8.34, –7.12).
60. Find the distance between the points (–12.5, –6.34) and (–3.24, 10.5).

The midpoint between two points (x_1, y_1) and (x_2, y_2) is given by

$$\left(\frac{x_1 + x_2}{2}, \frac{y_1 + y_2}{2}\right).$$

61. Find the midpoint between the two points in problem 59.
62. Find the midpoint between the two points in problem 60.
63. Evaluate $\left(1 + \frac{1}{n}\right)^n$ for n = 1, n = 2, n = 5, n = 10, n = 100, n = 1000, and n = 10000.
    (a) Do these number seem to get close to a particular value as n gets large?
    (b) If so, what does this value appear to be?
    (c) How large does n need to be in order for the successive values to differ by no more than 0.01 units.

Chapter 2

Functions and Graphing

2.1 Functions and Their Evaluation

(Washington, Sections 3.1 and 3.2)

A real function is a relationship between two sets of real numbers such that for every value from the first set of numbers there is exactly one value of the second set of numbers that relates to it. Variables are used to represent elements in these sets of numbers. A variable representing a number in the first set is referred to as the **independent variable** and a variable representing a number in the second set is referred to as the **dependent variable**. Generally, we will let x represent a number in the first set or independent variable and y represent the corresponding number in the second set or dependent variable. The relationship between the two sets of numbers can often be expressed in the form of an equation relating x and y. For example, $y = x^2 + 2$ states that each number in the first set is squared and added to 2 to obtain a number in the second set .

In mathematics, functional notation is often used to describe a function. The relationship $y = x^2 + 2$ is expressed in functional notation as $f(x) = x^2 + 2$. The evaluation of a function when x = k is indicated as f(k) and means to substitute the value of k for x in the functional expression and evaluate. For example, if we wish to evaluate the function $y = x^2 + 2$ at x = 3, we find f(3). This means to substitute 3 in for x everywhere x appears on the right side of the equation. Thus, $f(3) = 3^2 + 2 = 11$.

We will consider the **domain** of a real function to be that collection of real numbers such that when the function is evaluated at any of these numbers a real number is obtained. The **range** of a function is that collection of real numbers obtained by evaluating the function for all real numbers in the domain.

In order to evaluate a function at different values of x, we first store the function in the calculator's memory.

Procedure C10. To store a function.

On the TI-82:
The TI-82 may store up to eight functions, using function names $Y_1, Y_2, Y_3, ... Y_8$.

Method I:
1. Enter inside of quotes the function to be stored.
2. Press the key STO▷
3. Press the key Y-VARS
4. Select 1:Function
5. Select a function name (This function will replace any previous function.)
6. Press ENTER

To see stored functions, press the key Y=.

Method II:
1. Press the key Y=
2. Enter the function to the right of a function name.
3. Press QUIT

To see a stored function, press the key Y=.

On the TI-85:
The TI-85 may store up to 99 functions using names that may contain up to eight characters. The first character of a name must be a letter.

Method I:
1. Enter the name of the function. (Names may be up to eight characters long.)
2. Press the key =
3. Enter the function.
4. Press ENTER

To see a stored function, press the key RCL, enter the name of the function, and press ENTER.

Method II:
(Stores functions under the names y1, y2, y3, ...)
1. Press the key GRAPH
2. Select: y(x)=
3. Enter the function after a function name.
4. Press QUIT

To see a stored function, press the key GRAPH, select y(x)= and press ENTER.

Note: Pressing CLEAR while the cursor is on the same line as the function will erase the function. Be sure to QUIT the function list before evaluating a function.

After storing a function in memory we can use the calculator to evaluate the function at one or more values of the independent variable.

Procedure C11. To evaluate a function.

Functions may be evaluated at a single value or at several values by entering the values in a list.

On the TI-82:

1. Obtain the function name
   (a) Press the key Y-VARS
   (b) Select 1:Function...
   (c) Select the name of the function.
2. After the function name enter, in parentheses, the x value or a list of x values at which the function is to be evaluated. (Example: $Y_1(4)$ or $Y_1(\{1,2,3,4\})$)
3. Press ENTER

The results will be given as a value or a corresponding list. The TI-82 also has the ability to create a table of values for a function by pressing the key TABLE. Parameters for the table are set by pressing the key TblSet.

In TblSet:
- TblMin is the value of the first x value in the table.
- ΔTbl is the difference between table values.

On the TI-85:

1. Obtain the function evalF
   (a) Press the key CALC
   (b) Select evalF
2. Enter either the function or the function name then a comma.
3. Enter the variable for which the function is being evaluated followed by a comma.

The function names y1, y2, etc. may be entered from the keyboard (y must be lower case) or may be selected by:
   (a) Pressing the key VARS
   (b) From menu, select EQU (after pressing MORE).
   (c) Select the variable name (may need page↓ or page↑).

4. Enter the value or a list of values at which the function is to be evaluated. (Example: evalF(y1, x, 4) or evalF(y1,x,{1,2,4})
5. Press ENTER

Note: In the evaluation of a function, instead of placing a list of values inside of braces {}, we may enter the list name. See Procedure C7 for entering lists.

The appearance of an error message when evaluating a function likely indicates that the value of x is not in the domain of the function.

Example 2.1: Evaluate the function $f(x) = \sqrt{x+3}$ for x values of 6, 2.5, –3, and –4. Are any of these values not in the domain of f(x)? Round off answers to 3 significant digits.

Solution:

1. Use Procedure C7 to store the function $\sqrt{x+3}$ and then Procedure C8 to evaluate at the x values.
2. The value of f(6) is 3.00
   The value of f(2.5) is 2.35.
   The value of f(–3) is 0.00.
   We find that –4 is not in the domain of the real function f(x). (Some calculators may display (0,1) indicating that the complex number, j, is the solution.)

Sometimes, it is helpful to enter a constant as an alpha character when storing a function in the calculator. The value for the constant is stored for the alpha character and may be changed by storing a new value for the alpha character. For example, to evaluate $\pi x^2 h$ for several values of x when h = 1.5 and also for h = 1.8, we would first store the function $\pi x^2 h$ for a function name, store 1.5 for h and evaluate the function, then store 1.8 for h and again evaluate the function.

Procedure C12. To store a constant for an alpha character.

1. Enter the value to be stored.
2. Press the key STO▷
   (STO▷ shows as → on screen.)
3. Enter the alpha character.
4. Press ENTER

(Example: 6.35 → H)

Example 2.2: Evaluate $F(x) = \pi x^2 h$ for x = 2.50, 2.60, and 2.70 cm when (a) h = 6.35 cm and when (b) h = 6.45 cm.

Solution:

1. Use Procedure C7 to store the function $\pi x^2 h$ for a function name.
2. Use Procedure C9 to store 6.35 for H.
3. Use Procedure C8 to evaluate the function for x = 2.50, 2.60, and 2.70. We find that F(2.50) = 125, F(2.60) = 135, and F(2.70) = 145.
4. Now, use Procedure C9 to store 6.45 for H.
5. Evaluate the function again for x = 2.50, 2.60, and 2.70. This time we obtain F(2.50) = 127, F(2.60) = 137, and F(2.70) = 148.

Exercise 2.1

Evaluate the functions at the given values of x. Tell which values of x are not in the domain of the function. Round off decimal numbers to 2 decimal places.

1. $f(x) = x^2 + x$ for $x = -3$
2. $g(x) = x^3 - 2x^2 - 4$ for $x = 2.5$
3. $g(x) = \dfrac{2}{x-2}$ for $x = 2.11$ and $x = 2$
4. $h(x) = \dfrac{3}{x+4}$ for $x = -4.1$ and -4
5. $f(x) = \dfrac{x-3}{x^2-4}$ for $x = -2, 0, 2.3, 3, \sqrt{2}$
6. $g(x) = \dfrac{x-5}{x^2-2}$ for $x = 5, 2, \sqrt{2}, 3, 0$
7. $h(x) = \sqrt{x^2 - 5}$ for $x = 3, 2, -4, 0, 5$
8. $f(x) = \sqrt{x^2 - 9}$ for $x = 5, -5, 4, -4, \sqrt{3}$
9. $g(x) = \sqrt{x^2 + 5}$ for $x = 2, 5, -2, 0, \sqrt{5}$
10. $h(x) = \sqrt{x^2 + \pi}$ for $x = 3, 4, -4, 0, \sqrt{\pi}$
11. $f(x) = |x - 3|$ for $x = 3, 3.1, -3.1, 5.2, 0$
12. $F(x) = |x + 2|$ for $x = 3.12, -2.15, 2.78, 1.32, -2$
13. $G(x) = [x - 1]$ for $x = 2.3, 1.5, -2.34, 1.99, 0.999$
14. $f(x) = [2 + x]$ for $x = 0, 2.001, -2.001, -1.999, 1.002$
15. Evaluate $S(x) = h - 9.8x^2$ for $x = 1.0, 1.2$, and 1.4 first when (a) $h = 19.5$, then when (b) $h = 17.5$.
16. Evaluate $g(t) = at^2 - a^2t$ for $t = -0.50, 0.50$, and 1.50 first when (a) $a = -21$, then when (b) $a = 18$.
17. Evaluate $y = mx + b$ for the x-values 3.00, 3.20, 3.40, and 3.60. First when (a) $m = 6.00$ and $b = -2.00$ and then when (b) $m = 7.00$ and $b = 3.00$.
18. Evaluate $y = a(x - 4.55)^2 - k$ for x values of 2.55, 3.55, 4.55, 5.55, and 6.55 first when (a) $a = 3.20$ and $k = 1.45$, then when (b) $a = -2.40$ and $k = -1.65$.
19. A baseball is thrown upward at a velocity of 62.0 feet per second. The function $h(t) = 62.0t - 16.0t^2 + 5.50$ gives the height of the baseball above ground level as a function of time t in seconds. Evaluate this function to find the height of the ball after 1.0 seconds, 1.875 seconds, and 3.75 seconds.
20. The electric power P, in watts, dissipated in a resistor R, in ohms, is given by the function $P = \dfrac{200R}{(100 + R)^2}$. Since $P = f(R)$, find $f(100)$, $f(110)$, and $f(120)$.

♦ 21. *(Washington, Exercises 3.2, #19)* A rocket burns up at a rate of 2.17 tons per minute after falling out of orbit into the atmosphere. If the rocket weighed 5874 tons before reentry, express its weight w as a function of the time t, in minutes, of reentry. Then evaluate the function to find the weight at 15.0 minutes, 21.0 minutes, and 25.0 minutes.

♦ 22. *(Washington, Exercises 3.2, #20)* A computer part costs $2.87 to produce and distribute. Express the profit p made by selling 129 of these parts as a function of the price of c dollars each. Then, evaluate the function to find the profit if the price is $4.78, $5.24, and $5.98.

♦ 23. *(Washington, Exercises 3.2, #23)* A company installs underground cable at a cost of $465.00 for the first 50.0 feet (or up to 50 feet) and $5.14 for each foot thereafter. Express the cost C as a function of the length l of underground cable if $l > 50$ feet. Then, evaluate the function to determine the cost if the distance is 65 feet, 97.2 feet, and 212.9 feet.

♦ 24. *(Washington, Exercises 3.2, #30)* A helicopter 117.5 feet from a person takes off vertically. Express the distance d from the person to the helicopter as a function of the height h of the helicopter. Then, evaluate the distance from the person when the helicopter is at a height of 22.3 feet, 52.6 feet, and 117.5 feet.

2.2 Graphing a Function.
(Washington, Section 3.4)

In order to graph a function by hand it is often necessary to evaluate the function at several values of the independent variable, plot these points on a graph, then connect the points with a smooth curve. The graphing calculator allows us to enter the function and let the calculator do the work of evaluating that function for many values of the independent variable. The points obtained are plotted on the calculator's screen much quicker than we could graph by hand. It is also possible to have more than one function appear on the graph at the same time.

Before proceeding, it is helpful to understand the procedure used by the graphing calculator to graph functions. On the graphing calculator, there are only so many dots or pixels that may be plotted. To graph a function the x-value of each pixel in the horizontal direction is determined by the calculator and the corresponding y-value calculated from a entered function. If the screen has 95 useful pixels in the horizontal direction, the calculator will determine 95 points for the function. The calculator will turn on, or plot, all of these points that lie on the screen and (in connected mode) will turn on points lying between each pair of points. The difference in values as we move from one pixel to another determines the precision to which we can determine values of points on the screen.

When graphing on the graphing calculator in rectangular coordinates the independent variable will be x and the dependent variable will be y. When entering a function into the calculator we use a special key for the independent variable x.

<u>Procedure G1</u>. To enter and graph functions.

On the TI-82:

1. Press the key Y=
   This gives a screen showing a function table containing names of the functions:
   $Y_1=$
   $Y_2=$
   ...
2. Enter a function after the name of a function. Use the key X,T,θ for the variable x.
3. To graph functions, press the key GRAPH.
   To return to the calculations screen, press QUIT.

On the TI-85:

1. Press the key GRAPH
2. Select y(x)=
   This gives a screen showing a function table containing names of the functions:
   y1=
   ...
3. Enter a function after the name of a function. Use the key
   x-VAR for the variable x.
4. To graph functions, press EXIT to exit the secondary menu appearing on the screen, then select GRAPH from the menu.
   To return to the calculations screen, press QUIT.

Note:
1. Functions entered by this procedure will remain in the calculator's memory until changed or erased or until the memory is cleared.
2. While a graph is on the screen, pressing the key: CLEAR will clear the screen. However, the graph is still in the calculator's memory and can be seen again by pressing GRAPH.
3. When a function has been entered, the equal sign for that function becomes highlighted indicating that the function is turned ON - that is, it is an active function and will be graphed if the GRAPH key is pressed. Only functions which are turned ON will be graphed.

   To turn a function ON or OFF, with the function table on the screen:

   On the TI-82:
   (a) Move the cursor so that it falls on top of the equal sign of the function we wish to turn ON or OFF.

   On the TI-85:
   (a) Move the cursor so it is on the same line as the function to be turned ON or OFF.

(b) Press the ENTER key. If the function was ON it will be turned OFF and if it was OFF it will be turned ON.

(b) From the secondary, select SELCT. If the function was ON it will be turned OFF and if it was OFF it will be turned ON.

There are times when we may find it is necessary to change or erase a function. This may be done using the cursor, insert and delete keys.

Procedure G2. To change or erase a function.

1. Display the function table on the screen (see Procedure G1).
2. (a) To change a function: Use the arrow keys to move the cursor to the desired location and make the changes by inserting, deleting, or changing the desired characters.
   (b) To erase a function: With the cursor on the same line as the function, press the key: CLEAR
3. Select GRAPH to graph the function or exit to the calculations screen by pressing QUIT.

The view of a graph that we see on the screen of the graphing calculator is called the **viewing rectangle**. At this point we will use what is called the standard viewing rectangle and defer discussion on other viewing rectangles.

Procedure G3. To obtain the standard viewing rectangle.

To obtain the standard viewing rectangle;

On the TI-82:
1. Press the key ZOOM
2. From the menu, select 6:ZStandard

On the TI-85:
1. Press the key GRAPH
2. From the menu, select ZOOM
3. From the secondary menu, select ZSTD.

For the standard viewing rectangle, the x-value at the left of the screen is −10 and at the right of the screen is 10 and the y-value at the bottom of the screen is −10 and at the top of the screen is 10. Each mark on the axes represents one unit.

We are now ready to graph a function. We can generally enter functions just as they appear. Be sure to group quantities by correctly using parentheses.

Example 2.2: Graph the function $y = x^2 - 4$.

Solution:
1. Use Procedure G3 to obtain the standard viewing rectangle.
2. Use Procedure G1 to enter and graph the function

$y = x^2 - 4$. The graph is in Figure 2.1. From the screen it appears that the graph crosses the x-axis at −2 and at 2 and crosses the y-axis at −4.

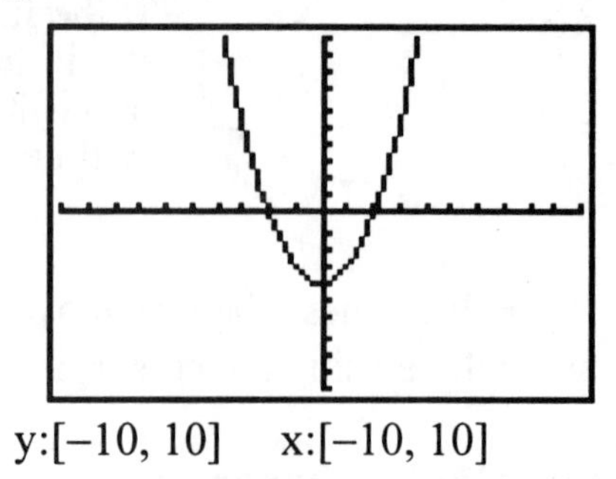

y:[−10, 10] x:[−10, 10]

Figure 2.1

Graphing calculators allow us to graph more than one function at a time. This feature allows us to compare graphs of different functions and to determine the points of intersection of intersecting graphs..

Example 2.3: Graph the functions $y = 2x$, $y = 2x - 2$, and $y = 2x - 4$. How do these graphs compare?

Solution:

1. Follow Procedure G1 to enter and graph the three functions. Enter 2x for the first function, 2x – 2 for the second function, and 2x – 4 for the third function. The resulting graph is in Figure 2.2.
2. The graphs are parallel straight lines which intersect the y-axis at different points (0, −2, and −4). We could conclude from this that in a function of the form $y = mx + b$, the b value represents where the graph crosses the y-axis.

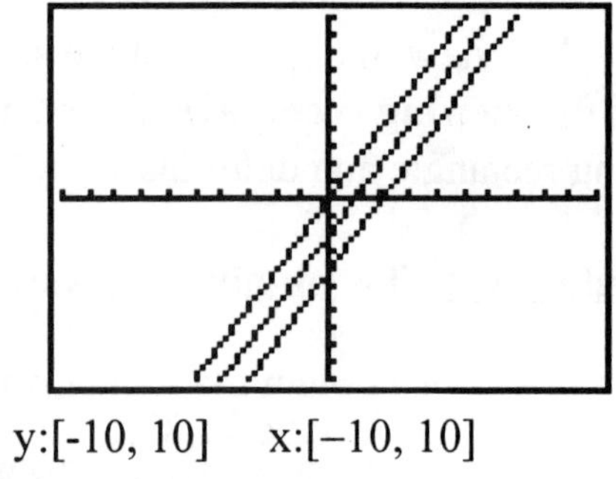

y:[-10, 10] x:[−10, 10]

Figure 2.2

There are certain points on a graph that we shall call important points. For certain applications these points may have special meanings. We shall consider the x-intercept, the y-intercept, the local maximum point and the local minimum point as important points. The x-value where the graph crosses the x-axis is called the **x-intercept** of the graph of the function. This value is also referred to as the **zero** of the function since when substituted into the function a value of zero is obtained. The y-value where the graph crosses the y-axis is called the **y-intercept** of the graph. If a graph is continuous, that is there are no breaks in the graph, then a point is a **local maximum point** if it is the highest point of all points in the immediate vicinity and if there exists points on either side of the point. Like wise, a point is a **local minimum point** if it is the lowest point of all points in the immediate vicinity and if there exists points on either side of the point. The graph of a function may have several x-intercepts, several local maximum points, and several local minimum points, but only one y-intercept.

The graph in Figure 2.3 has 2 local maximum points (at about $x = -4.5$ and $x = 8$), one local minimum point (at about $x = 1$) and four x-intercepts (at about $x = -7.4$, $x = -1$, and $x = 3$). One of the x-intercepts is off the right of the graph. When giving a local maximum point or a local minimum point, we give the coordinates of the point as an ordered pair. If a local maximum point occurs at $x = -4.5$ and $y = 3$, we state the local maximum point as $(-4.5, 3)$.

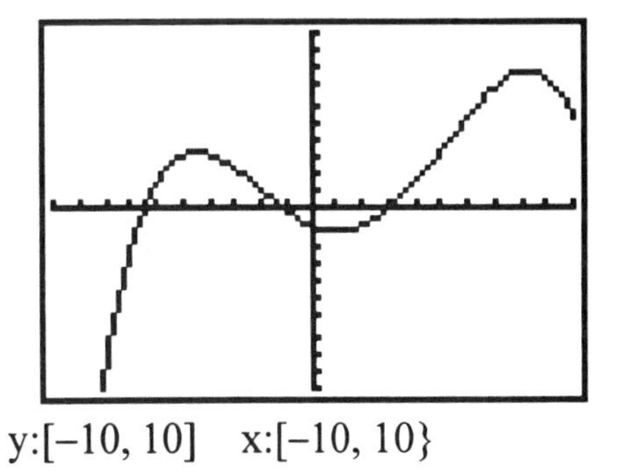

y:[−10, 10] x:[−10, 10}

Figure 2.3

It may be helpful when determining points on a graph to use the trace function on the calculator. With the trace function, the cursor moves along the graph and the x and y coordinates of points on the graph are given at the bottom of the screen. One press of the left or right cursor key moves the cursor one pixel in the horizontal direction and one press of the top or bottom cursor key moves the cursor one pixel in the vertical direction. Thus, by moving the cursor to a particular point on the graph, the coordinates of that point are known. Caution is urged in reading the coordinates since the precision of these points depends on the scale of the x- and y-axes.

Procedure G4. To use the trace function and find points on a graph.

1. Select the trace function:

On the TI-82:	On the TI-85:
Press the key TRACE	With the graph menu on the screen, select TRACE.

2. As the right and left arrow keys are pressed, the cursor moves along the graph and, at the same time, the coordinates of the cursor are given on the bottom of the screen.
3. If more than one graph is on the screen, pressing the up or down cursor keys will cause the cursor to jump from one graph to another. The number of the function will appear in the upper right of the screen.
4. If the graph goes off the top or bottom of the screen the cursor will continue to give coordinates of points on the graph.
5. If the cursor is moved off the right or left of the screen the graph scrolls (moves) right or left to keep the cursor on the screen.

Note: If the cursor is moved by using the cursor keys without first pressing the trace key, then the cursor can be moved to any point on the screen. The coordinates of this point will be given at the bottom of the screen.

◆ Example 2.4: *(Washington, Section 3.4, Example 3)* Graph the function $y = 2x^2 - 4$. Approximate the x- and y-intercepts and any local maximum or minimum points. Round answers to one-tenth of a unit.

Solution:

1. Use Procedure G1 to enter and graph the function $y = 2x^2 - 4$ on the standard viewing rectangle.
2. Use the trace function as in Procedure G4 to help estimate the intercepts and maximum point of the graph. The graph with the cursor at minimum point is shown in Figure 2.4.
3. From the graph, we approximate the x-intercepts, –1.5 and 1.5, the y-intercept, –4.0, and the maximum point (0.0, –4.0).

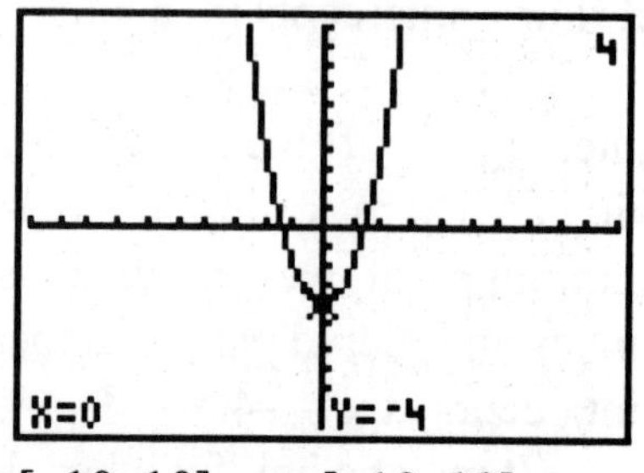

y:[–10, 10] x:[–10, 10]

Figure 2.4

Exercise 2.2

In Exercises 1-16, graph the given function on a graphing calculator using the standard viewing rectangle. In each case, make a sketch of the graph on your paper and then from the graph, use the cursor to estimate (a) the x-intercepts (zeros of the function), (b) the y-intercept, and (c) the coordinates of any local maximum and local minimum points on the graph. Round answers to one decimal place.

1. $y = 3.5x - 6.3$
2. $y = 0.47x + 1.5$
3. $f(x) = -2.4x + 9.7$
4. $f(x) = -3.5x - 8.9$
5. $y = x^2 - 7$
6. $y = 9.4 - 0.5x^2$
7. $f(x) = x^2 - 3x + 5$
8. $f(x) = -2x^2 + 19x - 42$
9. $y = x^3 - 7x$
10. $y = x^3 - 5x + 3$
11. $y = \frac{3}{x}$
12. $y = \frac{4}{x+4}$

13. $y = \dfrac{2x - 7}{x - 5}$
14. $y = \dfrac{5}{x^2 - 4}$
15. $y = \sqrt{2 - x}$
16. $y = \sqrt{18 - x^2}$
17. Graph $y = x - 2$, $y = x - 4$, and $y = x - 6$ on the same axes. What type of graphs are obtained? How do they compare?
18. Graph $y = x + 1$, $y = 2x + 1$ and $y = 4x + 1$ on the same axes. What type of graphs are obtained? How do they compare?
19. Graph several functions of the form $y = mx + b$. To do this replace m and b by several values of your choosing. (b should be between -10 and 10)
    (a) What kind of graph is always obtained?
    (b) What happens for positive values of m as m is increased?
    (c) What happens to the graph for negative values of m as m becomes a larger negative number?
    (d) What happens if $m = 0$?
    (e) What happens as b is change?
    (f) What happens if $b = 0$?
20. Graph several functions of the form $y = |mx + b|$. To do this replace m and b by several values of your choosing. (b should be between -10 and 10)
    (a) What kind of graph is always obtained?
    (b) What happens for positive values of m as m is increased?
    (c) What happens to the graph for various values of b?
    (d) What happens if $m = 0$? If $b = 0$?

♦ 21. *(Washington, Exercises 3.4, #33)* A force F (in lb) stretches a spring x inches according to the function $F = 3.78x$. Plot F as a function of x. Use the trace function to estimate the force when $x = 1.43$ inches and estimate the distance when $F = 7.8$ lb.

♦ 22. *(Washington, Exercises 3.4, #36)* The resistance R, in ohms, of a resistor as a function of the temperature R, in degrees Celsius, is given by $R = 4.5(1 + 0.082T)$. Plot R as a function of T. Use the trace function to estimate R when T = 3.65 degrees and estimate T when $R = 6.55$ ohms.

There are certain functions that we will call **Basic Functions**. Mental "snapshots" should be taken of these functions. Their shapes, intercepts, symmetry, and relationship to the x- and y-axes should be memorized. In Exercises 23-31 you are asked to graph these Basic Functions. In each case give the x- and y-intercepts, the values of $f(-1)$ and $f(1)$ and take a mental snapshot of the graphs. Estimate all values to one decimal place.

23. Constant function y = c. Where c is any constant.
24. Linear function: y = x. (y = x is often referred to as the identity function.)
25. Square function: $y = x^2$
26. Square root function: $y = \sqrt{x}$
27. Cube function; $y = x^3$
28. Cube root function: $y = \sqrt[3]{x}$
29. Reciprocal function: $y = \frac{1}{x}$
30. Absolute value function: $y = |x|$
31. Greatest integer function: y = [x]
    (On your calculator this is entered as Int(x).)

2.3 Changing Scales: Obtaining a Better Picture. *(Washington, Section 3.4)*

Often when graphing a function with a graphing calculator, important sections of the graph are off screen or parts of the graph may be too small to readily tell what the graph looks like. In these cases, it is necessary to adjust the viewing rectangle to allow us to obtain a better view of the graph. In Section 2.2 we used the standard viewing rectangle and in this section we introduce other viewing rectangles.

The **viewing rectangle** is identified by giving values for the quantities: Xmin, Xmax, Xscl, Ymin, Ymax, and Yscl. These are defined as:

Xmin is the x-value at the left of the screen.
Xmax is the x-value at the right of the screen.
Xscl is the scale value on the x-axis and indicates the number of units between marks on the x-axis.
Ymin is the y-value at the left of the screen.
Ymax is the y-value at the right of the screen.
Yscl is the scale value on the y-axis and indicates the number of units between marks on the y-axis.

<u>Procedure G5</u>. To see or change viewing rectangle.

1. To see values for the viewing rectangle on the screen.

On the TI-82:	On the TI-85:
Press the key WINDOW	With the graph menu on the screen, select RANGE

2. To change viewing rectangle values:
   Make sure the cursor is on the quantity to be changed and enter a

new value for that quantity. To keep a value and not change it, either (a) just press the key ENTER or (b) use the cursor keys to move the cursor to a value to be changed.

3. To see the graph, after the new values have been entered, select GRAPH
   To return to the calculations screen, press the key: QUIT

Note: The standard viewing rectangle has the following values:

Xmin = −10	Ymin = −10
Xmax = 10	Ymax = 10
Xscl = 1	Yscl = 1

When a function is graphed, all x-values that correspond to points on the graph are in the domain of the function and all y-values that correspond to points on the graph are in the range of the function. The viewing rectangle on the graphing calculator defines a **restricted domain** and a **restricted range** for the function(s) under consideration and generally does not give a view of the whole graph. Since the viewing rectangle does not normally show the whole graph, the question arises as to how much of the graph we should try to view. We will normally try to view that part of the whole graph that contains the important points including intercepts of the graph with the coordinate axes and all local maximum and local minimum points.

Example 2.5: Graph the function
$y = x^3 - 20x + 10$ and adjust
the viewing rectangle to obtain a complete graph.

Solution:

1. Use procedure G1 to enter and graph the function
   $y = x^3 - 20x + 10$.
2. The graph appears in Figure 2.5. This does not give a very good view of the graph. It would be better if the y-scale covered a larger range of values.
3. Use Procedure G5 to change the viewing rectangle to get a more complete graph on the screen, then use the trace function to help estimate the minimum and maximum y values.
   (See Procedure G4).
4. It turns out that a good viewing rectangle has x-values from −10 to 10 and y-values from −50 to 50 with Yscl = 10.

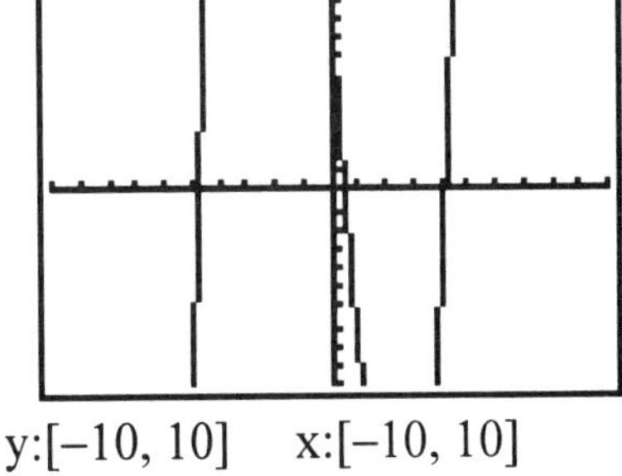

y:[−10, 10] x:[−10, 10]

Figure 2.5

Make these changes. The graph in Figure 2.6 gives a much better view.

In changing the scales for the axes, Xmin must be smaller than Xmax and Ymin must be smaller than Ymax. If either Xscl or Yscl is 0, no marks will appear on that axis.

The domain and range of a function may be estimated from the graph of the function. Each x-value for which there exists a point on the graph is in the domain of the function and each y-value that corresponds to a point on the graph is in the range of the function.

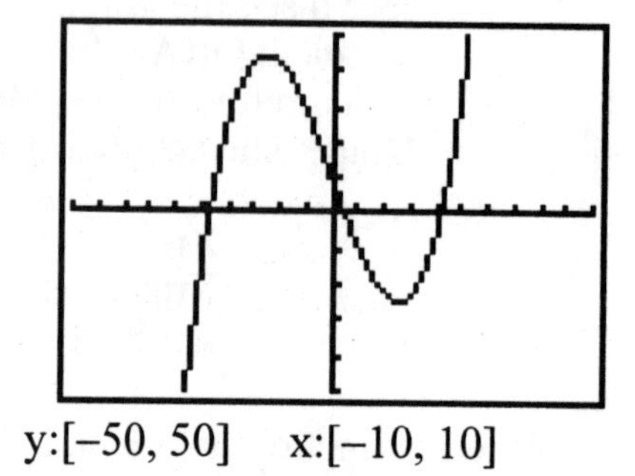

y:[−50, 50] x:[−10, 10]

Figure 2.6

Caution needs to be exercised when stating which values are in the domain or range of a function. On the standard viewing rectangle, the graph of the function $f(x) = \frac{x^2 - 1}{x - 1}$ is an apparent straight line. However, we can not evaluate this function at $x = 1$, since division by zero is not allowed. Thus, $x = 1$ is not in the domain and a point cannot exist on the graph of the function at $x = 1$. If this function is graphed using the decimal viewing rectangle (see Procedure G6), a hole appears in the graph at $x = 1$. This function is said to be discontinuous at $x = 1$. Functions are **discontinuous** at x-values that cause division by zero or at x-values where the function is not a real number. The graph of a discontinuous function does not exist at the point where the function is discontinuous.

It is also important to be able to determine the intervals on which a function is increasing or decreasing. If, on an interval of a graph of a function, the y-values always get larger as x-values increase, then the function is said to be **increasing** on that interval. If on an interval, the y-values always become smaller as the x-values increase, then the function is said to be **decreasing** on that interval. In an intuitive sense, this means that if the graph goes uphill as we look across the graph from left to right, the function is increasing. If the graph goes downhill, the graph is decreasing.

On occasion it is helpful to return to a standard viewing rectangle or to obtain other special viewing rectangles. Since the screen is a rectangle (width = 1.5 times height) and not a square, the length of one unit on the x-axis and the length of one unit on the y-axis are not the same when the minimum and maximum values of x and y are the same. If a circle is graphed using the standard viewing rectangle, it would not appear as a circle but as an ellipse. To remedy this, we can use a **square viewing rectangle** on which the x- and y-axes have the same scale distance for each unit.

Procedure G6: To obtain preset viewing rectangles.

1. Bring the zoom menu to the screen.

On the TI-82:	On the TI-85:
Press the key ZOOM	With the graph menu on the screen, select ZOOM

2. Select the desired preset viewing rectangle. (The left hand column gives the selection from the ZOOM menu for the TI-82 and the right hand column for the TI-85.)
   (a) For the standard viewing rectangle:
   Xmin = –10, Xmax = 10, Xscl = 1,
   Ymin = –10, Ymax = 10, Yscl = 1
   Select: 6:ZStandard ZSTD
   (b) For the square viewing rectangle:
   The square viewing rectangle keeps the y-scale the same and adjusts the x-scale so that one unit on x-axis equals one unit on y-axis. For a square viewing rectangle the ratio of y-axis to x-axis is about 2:3.
   Select: 5:ZSquare ZSQR (after pressing MORE)
   (c) For the decimal viewing rectangle:
   The decimal viewing rectangle makes each movement of the cursor (one pixel) equivalent to one-tenth of a unit
   Select: 4:ZDecimal ZDECM (after pressing MORE)
   (d) For the integer viewing rectangle:
   The integer viewing rectangle makes each movement of the cursor (one pixel) equivalent to one unit. After selecting the integer viewing rectangle, move the cursor to the point that is to be located at the center of the screen.
   Select: 8:ZInteger ZINT (after pressing MORE)
   Then press ENTER.
   Note: To have one movement of the cursor differ by k units set:
   Xmin = –94k/2 Xmin = –126k/2
   Xmax = 94k/2 Xmax = 126k/2
   (e) For the trigonometric viewing rectangle:
   The trigonometric viewing rectangle sets the x-axis up in terms of π or degrees (depending on mode)and sets Xscl = $\frac{\pi}{2}$ (or 90°).
   Select: 7:ZTrig ZTRIG

In Example 2.5, we did not see enough of the graph using the standard viewing rectangle. It is also sometimes necessary to enlarge a portion of the screen to better view the graph.

♦ Example 2.6: *(Washington, Section 3.4, Example 3)* Graph the function $y = x - x^2$, then from the graph
(a) estimate the x-intercepts of the graph to one decimal place,
(b) determine the range and domain of this function, and

(c) determine the intervals on which the function is increasing and on which intervals the function is decreasing.

Solution:

1. Graph the function $y = x - x^2$ on the standard viewing rectangle. The graph is in Figure 2.7. From this graph it is difficult to estimate where the graph crosses the x-axis.
2. Next, we change the scale to obtain a better view of the graph. Use Procedure G5 to set the scale values of Xmin = –1, Xmax = 3, Xscl = 1, Ymin = –1, Ymax = 1, Yscl = 1. This will give a restricted domain of –1 to 3 and a restricted range of –1 to 1.
4. Graph the function again. This graph appears in Figure 2.8. We now answer the questions.
   (a) We estimate the x-intercepts as being about 0.0 and 1.0. The trace function may be used as an aid to determine these values.

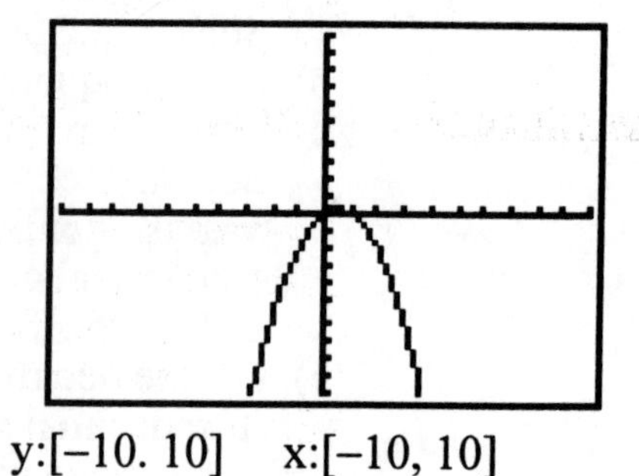
y:[–10. 10] x:[–10, 10]

Figure 2.7

   (b) From the graph and the function, it appears that there are no places where the graph is not continuous and each x-value corresponds to a point on the graph. We conclude that the domain of this function is all real numbers. The highest point on the graph occurs at $y = 0.25$ and all y-values below 0.25 appear to correspond to points on the graph. We conclude that the range is all real numbers y such that $y \leq 0.25$.

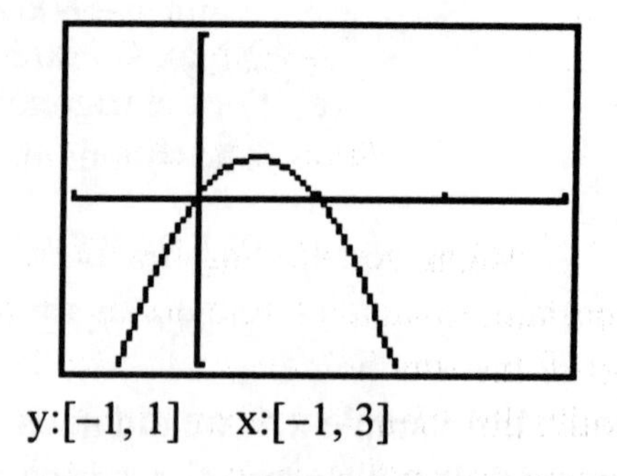
y:[–1, 1] x:[–1, 3]

Figure 2.8

   (c) From the graph we determine that the function decreases for $x > 0.5$ and increases for $x < 0.5$.

A type of function that we graph somewhat differently is a piecewise function. **Piecewise functions** are functions which are expressed in terms of two or more functions, each defined on specified intervals of the domain. To graph these functions on the graphing calculator it is necessary to place the specified interval in parentheses following the given function. The interval is identified using the relation symbols $=, \neq, >, \geq, <,$ and $\leq$.

Procedure C13. To obtain relation symbols ($=, \neq, >, \geq, <,$ and $\leq$).

1. Press the key: TEST
2. Select the desired symbol from the menu.

Procedure G7. To graph functions on an interval.

1. To graph a function on the interval $x < a$ or on the interval $x \leq a$ for some constant a:
   In the function table after the equal sign, enter the function f(x) followed by $(x<a)$ or $(x\leq a)$.
   (Example: To graph $y = x^2$ on the interval $x<2$, enter for the function: $x^2(x<2)$)
2. To graph a function on the interval $a < x < b$ or on the interval $a \leq x \leq b$ for some constants a and b:
   In the function table after the equal sign, enter the function f(x) followed by $(x>a)(x<b)$ or $(x\leq a)(x\leq b)$.
   (Example: To graph $y = x^2$ on the interval $-3\leq x\leq 2$, enter for the function: $x^2(x\leq -3)(x\leq 2)$)
3. To graph a function on the interval $x > a$ or on the interval $x \geq a$ for some constant a:
   In the function table after the equal sign, enter the function f(x) followed by $(x>a)$ or $(x\geq a)$.
   (Example: To graph $y = x - 5$ on the interval $x>2$, enter for the function: $(x-5)(x>2)$)

Note: The above forms may be combined by writing their sum.

When considering functional values or the graph of a piecewise function, it is important to watch the intervals on which the function is defined. The graphs of each part of the function only exists on the interval on which that part is defined. Care should also be taken when graphing with the calculator in connected mode since lines may be drawn that are not part of the actual graph..

♦ Example 2.7: *(Washington, Section 3.4, Example 7)* For the function, f(x), (a) give the graph, (b) give the x- and y-intercepts and (c) give the domain and range.

$$f(x) = \begin{cases} 2x+1 \text{ if } x \le 1 \\ 6-x^2 \text{ if } 1 < x \end{cases}$$

Solution:

1. Writing the f(x) this way indicates that $y = 2x + 1$, but only for values in the domain that are less than or equal to 1 and $y = 6 - x^2$ for domain values that are greater than 1.
2. Use Procedure G7 to graph this function using $2x + 1$ as one function and $6 - x^2$ as a second function. On the standard viewing rectangle the graph appears as in Figure 2.9.
3. Answering the questions:
   (b) There are two x-intercepts, $x = -0.5$ and $x = 2.5$. The y-intercept is 1.0. When using the trace function make sure that the cursor is on the correct curve. Notice that the line connecting the curves at $x = 1$ is not part of the graph.
   (c) The graph exists for all x-values and the domain is all real numbers. From the graph we see that only y-values less than 5 correspond to points on the graph. The range is all real numbers y such that $y < 5$.

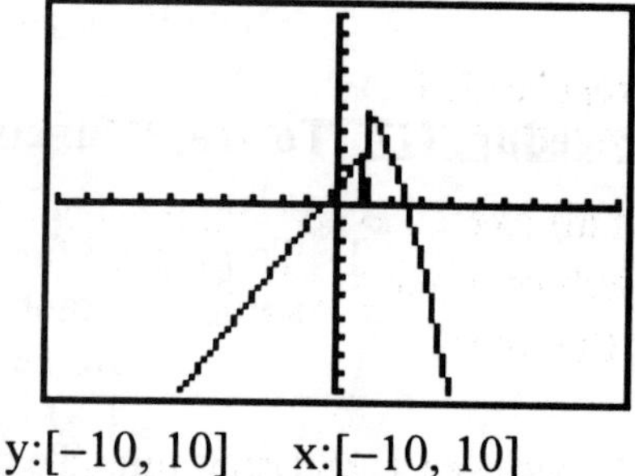

y:[−10, 10] x:[−10, 10]

Figure 2.9

When graphing discontinuous functions, that is functions which have breaks in the graph, lines sometimes appear on the graph that are not really part of the graph. With discontinuous functions it can be helpful to graph in dot mode. When the calculator is in **dot mode** individual points of the graph are plotted that correspond to each pixel on the x-axis. The points are not connected.

Procedure G8. Changing graphing modes.

1. To change modes:

On the TI-82:	On the TI-85:
Press the key MODE	With the graph menu on the screen, select FORMT (after pressing MORE)

Then, use the cursor keys to highlight the desired option.

2. To draw graphs connected or with dots

Connected for to connect plotted points with a line. Dot to just plot individual points.	DrawLine to connect plotted points with lines and DrawDot to just plot individual points.

3. To draw graphs simultaneously or sequentially

Sequential to graph each function in sequence or Simul to graph all selected functions at the same time.	SeqG to graph each function in sequence or SimulG to graph all selected functions at the same time..

4. Highlight the desired option in each line then press ENTER.
5. To return to the calculations screen, press CLEAR or QUIT.

Exercise 2.3

On all exercises, obtain a graph showing all important points and draw a sketch of the graph on your paper indicating values for Xmin, Xmax, Ymin, and Ymax at the ends of the axes.

In Exercises 1-18, from the graph estimate (a) the x-intercepts, (b) the y-intercepts, (c) the coordinates of the local maximum and local minimum points, (d) give the domain and range of each function, and (e) give the values of x for which the function is increasing and for which it is decreasing. Estimate answers to two significant digits - do not try to obtain a high degree of precision.

1. $y = x^2 - 5x - 6$
2. $y = 11 - 12x^2 - x^3$
3. $y = 252 - x^2$
4. $y = x^3 - 1200$
5. $y = x^3 - 66x$
6. $y = x^3 - 5x^2 + 6.25x$
7. $y = \dfrac{4x - 3}{3 - x}$
8. $y = \dfrac{x^2 - 5}{3x - 7}$
9. $y = \dfrac{x^2 - 16}{4 - x}$
10. $y = \dfrac{2x^2 + 5x - 3}{x + 3}$
11. $y = \sqrt{15 - x^2}$
12. $y = -\sqrt{20 - 14x - x^2}$

13. $y = \sqrt{95 - x^2 + 9x - 2}$
14. $y = 4x^3 - 44x^2 + 100x - 64$
15. $y = x + \dfrac{3}{x}$
16. $y = x^2 + \dfrac{3}{x^2}$
17. $y = |2.4x - 7.5|$
18. $y = |x^2 - 3x - 2|$
19. $f(x) = \begin{cases} 2 + x \text{ if } x \le 2 \\ (x-2)^2 \text{ if } x > 2 \end{cases}$
20. $g(x) = \begin{cases} \dfrac{1}{x-1} \text{ if } x < 1 \\ \sqrt{x-1} \text{ if } x \ge 1 \end{cases}$

Do exercises 21-24 as indicated. Estimate values to two significant digits.

♦ 21. *(Washington, Exercises 3.4, #37)* The consumption c of fuel, in liters per hour (L/h), of a certain engine is determined as a function of the number r, in revolutions per minute (r/min) of the engine to be $c = 0.0132r + 3.85$. This formula is valid from 500 r/min to 2000 r/min. Plot c as a function of r. Use the trace function to estimate the consumption when $r = 1100$ r/min and to estimate the revolutions per minute when $c = 26$ L/h.

♦ 22. *(Washington, Exercises 3.4, #38)* The profit P, in dollars, a manufacturer makes in producing x units is given by $P = 2.85x - 90.0$. Graph P as a function of x for x. (a) For what values of x is P positive? What is the significance of this? (b) What is the profit when $x = 20$ units? When $x = 59$ units? (c) How many units must the manufacturer make in order to have a profit that exceeds $100?

♦ 23. *(Washington, Exercises 3.4, #41)* The maximum speed v, in miles per hour, at which a car can safely travel around a circular turn of radius r, in feet, is given by $r = 0.42v^2$. Graph r as a function of v. Allow for speed up to 75 miles per hour. (a) What needs to be the radius for a car traveling 38 miles per hour? For a car traveling 75 miles per hour? (b) At what speed can a car travel around a curve of radius 1075 feet? Around a curve of radius 249 feet.

♦ 24. *(Washington, Exercises 3.4, #42)* The height h, in meters, of a rocket as a function of the time t, in seconds, is given by the function $h = 1449t - 4.9t^2$. Graph h as a function of t. Assume level terrain. (a) What is the greatest height obtained by the rocket? (b) For what values of time is the rocket going up (the height increasing)? (c) After what period of time does the rocket hit the ground?

In Exercises 25-30, give (a) the domain and range of the given basic functions, (b) the intervals on which each function is increasing or decreasing, and (c) the y-values for $x = -1, 0$, and 1.

25. $y = x$
26. $y = x^2$
27. $y = \sqrt{x}$
28. $y = \frac{1}{x}$
29. $y = |x|$
30. $y = [\,x\,]$

In Exercises 31 and 32, first graph the functions on the same standard viewing rectangle, then on the same square viewing rectangle. Connect the graph (where points are missing). (a) What type of graph is obtained by combining the two functions? (b) What are the x- and y-intercepts? (c) What are the domains and ranges of each function?

31. $y_1 = \sqrt{25 - x^2}$ and $y_2 = -\sqrt{25 - x^2}$
32. $y_1 = \sqrt{50 - x^2}$ and $y_2 = -\sqrt{50 - x^2}$

2.4 Zoom: Finding Special Points / Solving Equations *(Washington, Section 3.5)*

In section 2.3, we changed the viewing rectangle to obtain a better view of the graph. In this section, we introduce a generally more efficient method of changing the viewing rectangle by making use of the zoom function of the graphing calculator. The **zoom function** allows us to zoom in or to zoom out on the graph in the region of interest.

Procedure G9. **To zoom in or out on a section of a graph.**

1. With a graph on the screen,
   On the TI-82:
   Press the key ZOOM

   On the TI-85:
   With the graph and graph menu on the screen, select ZOOM

2. From the menu:
   To zoom in, select 2:Zoom In
   or
   To zoom in, select 3:Zoom Out

   From the secondary menu,
   To zoom in, select ZIN
   or
   To zoom out, select ZOUT

3. The cursor keys may be used to move the cursor near the point of interest. After zooming, the point at the location of the cursor will be near the center of the screen.
4. Press the key: ENTER
5. After zooming, repeated zooms may be made (if no other key is pressed in the meantime) by moving the cursor near the desired point and pressing ENTER.

Note: It is a good idea to use the TRACE function together with ZOOM.

The magnification determined of the zoom operation is determined by the **zoom factors**. These factors are preset on new calculators but may be changed to allow a greater magnification. If the zoom factors are set too large, it is possible that the portion of the graph that is of interest may not appear and only a blank screen will be obtained.

Procedure G10. To change zoom factors

1. Obtain the zoom menu:

On the TI-82:	On the TI-85:
Press the key ZOOM	With the graph menu on the screen, select ZOOM

2. Then

Highlight MEMORY in the top row and select 4:SetFactors.	From the secondary menu select ZFACT (after pressing MORE twice).

3. Change the factors as desired.
   There are factors for magnification in both the x- and y-directions. The preset factors are 4 in both directions. Make changes to XFact for the magnification in the x-direction and to YFact for magnification in the y-direction.
4. To exit this screen press the key: QUIT

The zoom function may be used to zoom in and find the coordinates of the point to any desired accuracy (limited only by the capability of the calculator). Example 2.8 uses the zoom function to find the x-intercepts.

Example 2.8: Graph the function $y = x^2 - 3x - 5$ and locate the x-intercepts to two decimal places.

Solution:

1. Start by graphing the function, $y = x^2 - 3x - 5$ in the standard viewing rectangle. The graph is a parabola that crosses the x-axis near -1 and near $+4$. We will first find the x-intercept near -1.

2. Use Procedure G10 to check the zoom factors. Set both zoom factors to 4.
3. Return to the graph and use the trace function to move the cursor to a point on the graph near the negative x-intercept. On the bottom of the screen are the x- and y-values that give the location of the cursor on the graph. With the cursor near x = −1, the screen appears as in Figure 2.10. (Your values may differ.)

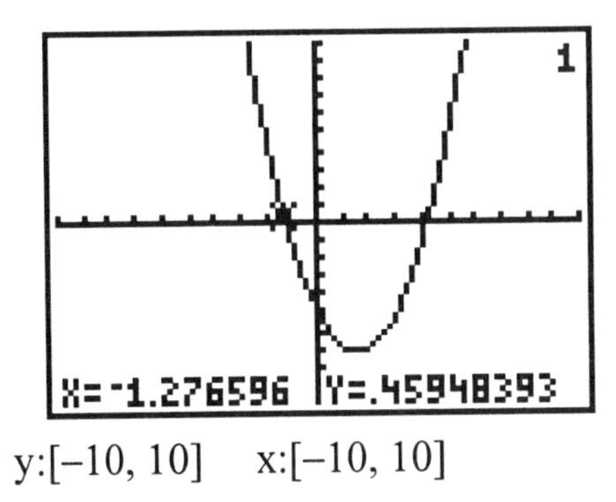

y:[−10, 10] x:[−10, 10]

Figure 2.10

4. Use Procedure G9 to zoom in.
5. Use the trace function and move the cursor near where the graph crosses the x-axis. The graph is given in Figure 2.11. At one point the cursor will be below the x-axis and at the next point to the left will be above the x-axis. This can also be determined by noting that for one point the y-value is negative and for another point the y-value is positive. The x-intercept must be between the two points. In this case, this means the solution must be between approximately x = −1.22 and x = −1.17.

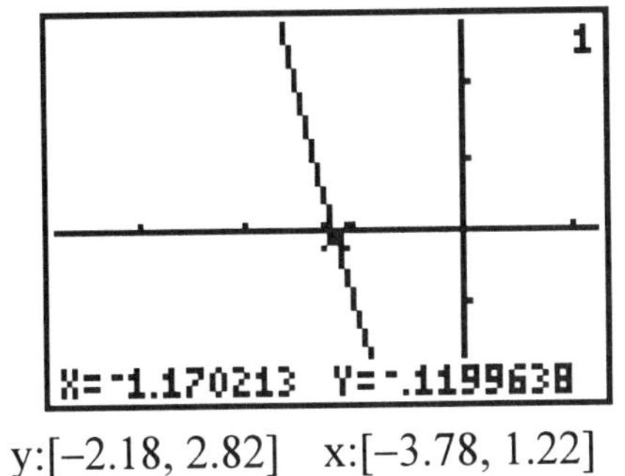

y:[−2.18, 2.82] x:[−3.78, 1.22]

Figure 2.11

6. Move the cursor to a point near the intersection point of the graph and the x-axis (where x = -1.27) and zoom in again (Procedure G9).
7. Repeat steps 5 and 6 until the x-values immediately on either side of the intersection point round off to the same two decimal place number. The x-intercept is −1.19.
8. To determine the other x-intercept, first return to the standard viewing rectangle and obtain the original graph.
9. Move the cursor to the positive x-intercept and repeat the procedure until the zero near x = 4 can be found to two decimal places. The x-intercept is 4.19.

The x-intercepts of $y = x^2 - 3x - 5$ are −1.19 and 4.19.

The location of the x-intercept in Example 2.6 involved repeating the three steps:

1. Move the cursor near the desired point.
2. Zoom in the point.
3. Use the TRACE function to help determine the point.

Note: Continue until the desired degree of precision is obtained.

In Example 2.6, we zoomed in on a section of the graph. In some cases, it may be desirable to zoom out in order to see more of the graph. Procedure G9 is used to zoom out on a section of the graph and may be repeated as necessary to obtain the desired results

An approximate solution to an equation can be found relatively quickly and to a high precision by using a graphing calculator. In application problems, it is often not necessary to obtain an exact answer, but to find an answer to a desired precision. To solve an equation of the form $f(x) = 0$, where $f(x)$ represents a function of x, first let $y = f(x)$ and graph the function. The solutions to the equation $f(x) = 0$ are the x-intercepts of the graph. Solutions to an equation are often referred to as **roots** of the equation.

It is important to note that the solutions (or roots) to the equation $f(x) = 0$, the zeros of the function $f(x)$, and the x-intercepts of the graph of $y = f(x)$ are all the same numerical values.

Procedure G11. Solving an equation in one variable.

1. Write the equation so that it is in the form $f(x) = 0$, let $y = f(x)$ and graph the function. Use a viewing rectangle so that the x-intercept of interest appears on the screen.
2. Determine a solution by:
   (a) Using Procedure G9 to zoom in on the x-intercept.
   or
   (b) Using the built in procedure:

   On the TI-82:
   (1) Press the key CALC
   (2) From the menu, select 2:root
   The words Lower Bound? appear.

   On the TI-85:
   (1) From the graph menu, select MATH (after pressing MORE).
   (2) From the secondary menu, select ROOT.

(3) Move the cursor near, but to the left of the intercept and press ENTER. The words Upper Bound? appear.
(4) Move the cursor near, but to the right of the intercept and press ENTER. The word Guess? appears. It is sufficient at this point to just press ENTER again. The root (x-intercept) will appear at the bottom of the screen.

(3) Move the cursor near the intercept and press ENTER. The root (x-intercept) will appear at the bottom of the screen.

♦ Example 2.9: *(Washington, Section 3.5, Exercise 3)* Solve the equation $x^2 - 2x = 1$. Find the answers to five decimal places.

Solution:

1. First place the equation in the form $f(x) = 0$ by subtracting 5x and 6 from both sides. This gives $x^2 - 2x - 1 = 0$. Let $y = x^2 - 2x - 1$.
2. Graph the function $y = x^2 - 2x - 1$ in the standard viewing rectangle. From the screen, it appears this graph crosses the x-axis just to the left of zero and between 2 and 3.
3. Use Procedure G11 to determine the negative root between -1 and 0, then determine the root between 2 and 3. The roots are found to be $x = -0.41421$ and $x = 2.41421$.

We may use a similar method to find the coordinates of local maximum and local minimum points to a high degree of precision.

Procedure G12. Finding maximum and minimum points.

1. Graph the function and adjust the viewing rectangle so that the desired local maximum or local minimum point is on the screen.
2. Determine the local maximum or local minimum point by:
   (a) Using Procedure G9 to zoom in on the point
   or

(b) Use the built-in procedure:

On the TI-82:

(1) Press the key CALC
(2) From the menu, select 3:minimum or 4:maximum. The words Lower Bound? appear.
(3) Move the cursor near, but to the left of the desired point and press ENTER. The words Upper Bound? appear.
(4) Move the cursor near, but to the right of the desired point and press ENTER. The word Guess? appears. It is sufficient at this point to just press ENTER again. The coordinates of the point will appear at the bottom of the screen.

On the TI-85:

(1) Adjust the viewing rectangle so that the point of interest is the highest point or lowest point on the screen.
(2) From the graph menu, select MATH (after pressing MORE).
(3) From the secondary menu, select FMIN for a local minimum or FMAX for a local maximum point (after pressing MORE).
(4) Move the cursor near the desired point and press ENTER. The coordinates of the point will appear at the bottom of the screen.

Note: LOWER and UPPER may be used to select lower and upper bounds on which the maximum and minimum values are to be found.

Example 2.10: Find the local maximum and local minimum points on the graph of $y = 3x^2 - 7x - 1$ to four decimal places, give the x-values for which the function is increasing, and give the domain and range of the function.

Solution:

1. Graph the function $y = 3x^2 - 7x - 1$ on the standard viewing rectangle. From the graph we see that the function gives a parabola with a local minimum point.
2. Use Procedure G12 to determine the coordinates of the local minimum point. The minimum point is located at (1.1667, –5.0833).
3. From the graph we see that the graph is increasing to the right of the minimum point. Thus, the function increases for $x > 1.1667$.
4. From the graph and the function, we see that every x-value corresponds to a point on the graph. Thus, the domain is all real numbers.

   Since the graph has a minimum point, and there are no points on the graph below this point, the range must be the y-values corresponding this point or points above this point. Thus, the range is $y \geq -5.0833$.

On occasion it is helpful to be able to evaluate a function to a high precision at a given value of x.

Procedure G13. Find the value of a function at a given value of x.

1. Graph the function and adjust the viewing rectangle so that the given x-value is within the restricted domain of the screen.
2. Determine the value of the function by:
   (a) Using Procedure G9 to zoom in on the point
   or
   (b) Use the built-in procedure:

   On the TI-82:
   (1) Press the key CALC
   (2) From the menu, select 1:value. The words Eval X= appear on the screen.
   (3) Enter the given x-value and press ENTER. The value of the function will appear at the bottom of the screen.

   On the TI-85:
   (1) From the graph menu, select EVAL (after pressing MORE twice).
   (2) The words Eval x = appear on the screen.
   (3) Enter the x-value and press ENTER. The value of the function will appear at the bottom of the screen.

Another method that can be used to zoom in on a portion of a graph is to create a box containing the area to be enlarged and have the calculator redraw the region inside the box to fill the screen. This method can be used in combination with other zooming methods.

Procedure G14. To zoom in using a box.

1. Obtain the zoom menu:
   On the TI-82:
   Press the key ZOOM

   On the TI-85:
   With the graph menu on the screen, select ZOOM
2. Then
   Select 1:ZBox

   From the secondary menu select BOX
3. Use the cursor keys to move the cursor to a location where one corner of the box is to be placed and press ENTER.
4. Use the cursor keys to move the cursor to the location for the opposite corner of the box and press ENTER. As the cursor keys are moved a box will be drawn and when ENTER is pressed, the area in the box is enlarged to fill the screen.

Exercise 2.4

Graph each of the functions in exercises 1-8 on your calculator. Find (a) the x-intercepts and (b) the coordinates of local maximum and local minimum points, and (c) the domain and range of each function. Find answers to three significant digits.

1. $y = x^2 - x - 24$
2. $y = x^3 + 8x$
3. $y = 7 - 6x + 3x^2 - 8x^3$
4. $y = \sqrt{11 - x^2}$
5. $y = \dfrac{3x - 5}{x - 4}$
6. $y = \dfrac{3x + 1}{5 - 2x}$

Solve each of the following equations by graphing on the graphing calculator. Evaluate solutions to three significant digits.

7. $7x - 6 = 0$
8. $x^2 - 37 = 0$
9. $x^2 - 5x = 5$
10. $x^3 - 15x = 0$
11. $5x - 3x^2 + x^3 = 9$
12. $x^3 + 2x^2 = 6$

Do each problem as indicated. Find answers to four significant digits.

13. A small business depreciates some equipment according to the equation $V = 15500 - 3200t$ were V is the value in dollars after t years. According to this formula, when will the equipment be fully depreciated (have a value of zero)? Determine the value when $t = 2.5$ years and when $t = 3.25$ years.
14. The height, h, (in feet) of a baseball thrown upward from a cliff is a function of time, t, (in seconds) according to the equation $h = 150 + 70t - 16t^2$. At what time will the height be zero? At what time will the ball be at its highest point? For what times will the ball be descending? How high will the ball be when $t = 2.5$ s and when $t = 4.75$ s.?

♦ 15. *(Washington, Exercises 3.5, #25)* In an electric circuit, the current i, in amperes, as a function of the voltage v is given by $i = 0.0127v - 0.0558$. Find v when $i = 0$. For what values of v is the current positive? What will the current be when $v = 10.0$ and when $v = 7.5$?

♦ 16. *(Washington, Exercises 3.5, #28)* The pressure loss P (in lb/in^2 per 100 ft) in a fire hose is given by $P = 0.000208Q^2 + 0.01342Q$, where Q is the rate of flow (in gal/min). Find Q when $P = 11.8\ lb/in^2$. Find P when $Q = 99.7$ and when $Q = 125.4$.

♦ 17. *(Washington, Exercises 3.5, #36)* A rectangular box is to be made from a sheet of sheet metal, which measures 14.0 inches by 10.0 inches, by cutting out equal squares of side x from each corner and bending up the sides. Find the value of x that gives the maximum volume for the box. (Hint: Find a function for the volume of the box, then find a local maximum point on the graph.)

♦ 18. *(Washington, Exercises 3.5, #32)* For the box of problem 17 determine the value of x which will give a volume of 95 cubic inches.

2.5 Introduction to Programming the Graphing Calculator

The graphing calculator has the ability to be programmed in much the same way that one would program a computer to perform a specific task. A **program** is a set of individual instructions to tell the calculator what tasks are to be performed. In a program, each step must be explicitly stated. The symbols and words used to state these instructions form what is called a programming language. Each step of the program is called a statement and is a single instruction for the program.

The language used for programming the graphing calculator is different from other languages used for programming computers with which the reader may be familiar. To program the graphing calculator, we make use of the various keys and other instructions that are accessed by selection of menu items on the calculator. Essentially any calculation or instruction that may be done directly may also be programmed into the calculator. The best way to see how to enter and execute (run) a program is to follow the steps in Example 2.11. First we list the various procedures needed for programming. To enter a new program, it is first necessary to enter programming mode. **Programming mode** allows the instructions of the program to be entered or modified.

Procedure P1. To enter a new program.

1. Press the key: PRGM
2. Enter programming mode:

On the TI-82:
Highlight NEW in the top row and select 1:Create New

On the TI-85:
Select EDIT.

3. After Name=, enter a name for the program. The name may be up to eight characters long. (Notice that the calculator is ready for alpha characters.) Then, press ENTER.
4. The word PROGRAM appears on the top row followed by the name of the program and the cursor is on the second row preceded by a colon (:). The calculator is now in programming mode and is ready for the first statement in the program.
5. Enter the steps of the program one line at a time, pressing ENTER after each step.
6. When finished entering the program, press QUIT to return to the calculations screen.

Once a program has been entered it may be executed, edited or erased. To **execute** a program means that each of the instructions in the program is carried out in order, one at a time. To **edit** a program means to make changes in the statements of the program.

Procedure P2. To execute, edit, or erase a program.

1. Press the key: PRGM
2. To execute a program:

   On the TI-82:
   With EXEC highlighted in the top row, select the name of the program to be executed. Then, press ENTER..

   On the TI-85:
   Select NAMES
   From the secondary menu, select the name of the program. Then, press ENTER.

   A program may be executed additional times by pressing ENTER if no other keys have been pressed in the meantime.
3. To edit a program:

   On the TI-82:
   Highlight EDIT in the top row and select the name of the program to be edited.

   On the TI-85:
   Select EDIT.
   From the secondary menu, select the name of the program to be edited.

   Make changes in the program as desired. The cursor keys may be used to move the cursor to any place in the program. The calculator is now in programming mode.
4. To erase a program:

   On the TI-82:
   Press the key MEM.
   Select 2:Delete...
   From the secondary menu, select 6:Prgm...

   On the TI-85:
   Press the key MEM
   Select DELET
   Select PRGM
   (after pressing MORE)

   Move the marker on the left until it is next to the name of the program to be erased and press ENTER. The program will be erased from memory.

As mentioned previously, a program consists of a series of explicit steps that tell the calculator what operations to perform. Quite often it is desirable to enter a value from the keyboard for a variable. This is referred to as inputting a value. The value entered is then stored for the specified variable. When the variable is later used in a calculation, the value which has been stored replaces the variable in the calculation.

Procedure P3. To input a value for a variable.

Entering a value for a variable uses the keyword Input.
The form of the statement is: **Input v** (v represents the variable name)

With the calculator in programming mode:

1. To access the word Input:

On the TI-82:	On the TI-85:
Press the key PRGM Highlight I/O in the top row. Select 1:Input.	From the menu, select I/O From the secondary menu, select Input

2. Enter the variable name and press ENTER.
   (Example: Input C)

When the program is executed, the word Input followed by a variable will cause a question mark to appear on the screen. Any number entered is stored for that variable and when the variable is later used in the program that value is used in the calculation.

Once a value for a variable has been entered, it is likely that it will be used to evaluate an expression. This is accomplished by entering a statement which gives the expression and stores the result to another variable.

Procedure P4. To evaluate an expression and store that value to a variable.

With the calculator in programming mode:

On the TI-82:	On the TI-85:
Enter the expression, press the key STO▷, and then, enter a variable name.	Enter the expression, press the key STO▷ and then, enter a variable name. or Enter a variable name, press the key =, and then enter the expression.

(Example: $\frac{9}{5}C + 32 \rightarrow F$)

After an expression has been evaluated and the results stored for a variable, the results need to be displayed. It is also possible to display words or string expressions on the screen. A **string expression** is a set of characters enclosed in quotes.

Procedure P5. To display quantities from a program.

Either the value for a variable or an alpha expression may be displayed by using the keyword Disp.
The form of the statement is: **Disp v** (v represents the variable name)
The quantity v may be a variable name, a number, or a set of characters enclosed inside of quotes. If v is a quantity enclosed in quotes, the quantity will be printed exactly as it appears.

With the calculator in programming mode:
1. To access the word Disp:

On the TI-82:	On the TI-85:
Press the key PRGM	From the menu, select I/O
Highlight I/O in the top row.	From the secondary menu,
Select 3:Disp.	select Disp

2. Enter the variable name and press ENTER.
   (Example: Disp F)

When the program is executed, the value stored for the variable or the expression will be displayed on the screen.

Now, we illustrate how a program is entered and executed. When programming, it a good idea to enter the basic steps needed, then, when we are certain the program works, go back and make improvements.

Example 2.11: Write a program to calculate the temperature in degrees Fahrenheit when a temperature in degrees Celsius is entered. Use the formula $F = (9/5)C + 32$. Then, evaluate F if $C = 100$ and if $C = -40$.

Solution:
1. Obtain the correct mode for entering a new program by using Procedure P1.
2. For the program name enter: CTOF
3. Use Procedure P3 to enter on the second line enter a statement to input a value for C. (Input C)
4. On the next line, use Procedure P4 enter the expression

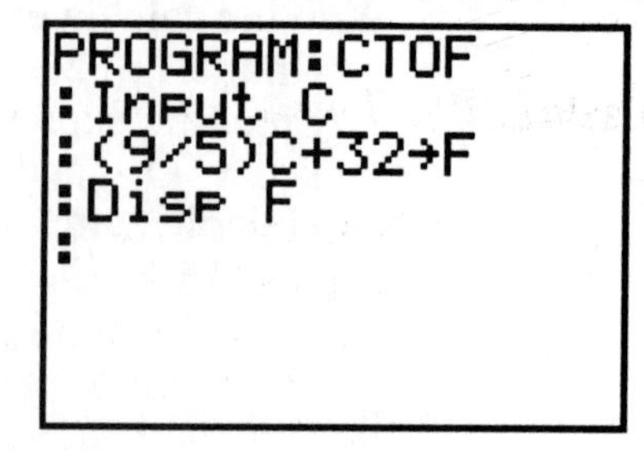

Figure 2.12

$\frac{9}{5}C + 32$ and store the resulting value for F. ((9/5)C + 32 →F)

5. On the next line use Procedure P5 to display the value for F (Disp F). This constitutes a brief, but complete program. The program appears in Figure 2.12.
6. Exit from the programming mode, then execute this program (see Procedure P2). The input statement will cause a question mark to appear on the screen. The program is requesting a numerical value to be entered for the variable C. Enter the number 100 and press ENTER. The corresponding value for F should appear at the right of the screen. The result is 212.
7. The program may be executed several more times by pressing the key ENTER if no other keys are pressed in the meantime. If another key has been pressed, use Procedure P2 to execute the program. Execute the program a second time and enter –40. The result should be –40. This is the temperature where Celsius and Fahrenheit temperatures are the same.

If errors occur when entering a program or if the program needs to be changed, the lines of the program may be edited by moving the cursor to the desired position and making the changes. (See Procedure P2.)

One difficulty with the program in Example 2.11 is that when the question mark appears on the screen, we may not be aware of the value desired. Likewise, when the value is printed out we may not be aware of its meaning. This is particularly true in large programs where several variables are used. This problem may be overcome by printing a word or two prior to the question mark or output. Example 2,.12 illustrates how we handle this. A word or two that appears before the question mark is called a **prompt**. Before considering the example, we consider how to add or delete instructions from a program.

Procedure P6. To add or delete line from a program.

With the calculator in programming mode:

1. To add a line:
   (a) To insert a line before a given line move the cursor to the first character of the line and to insert a line after a given line move the character to the last character in the given line.
   (b) Press the key INS, then press ENTER This will create a blank line either before or after the given line
   (c) Make sure the cursor is in the desired line. An instruction may now be entered.

2. To delete a line
   (a) Move the cursor to the desired line
   (b) Press the key CLEAR, then press DEL.

Example 2.12: For the program converting Celsius temperatures to Fahrenheit, have a prompt displayed before the question mark and identify the output.

Solution:

1. Use Procedure P2 to edit the program CTOF.
2. Use Procedure P6 to insert a blank line before the line Input C.
3. To display a message that will indicate what value is to be entered, use Procedure P5 to insert the instruction to display the string "Enter C" (Disp "ENTER C")
4. Next, we desire to identify the output of the program. Use Procedure P6 to insert a blank line before the line Disp F.
5. Now, as in step 3 enter the instruction to display "Degrees F". (Disp "DEGREES F ")
6. Execute the program. Try several values for C. The complete program appears in Figure 2.13.

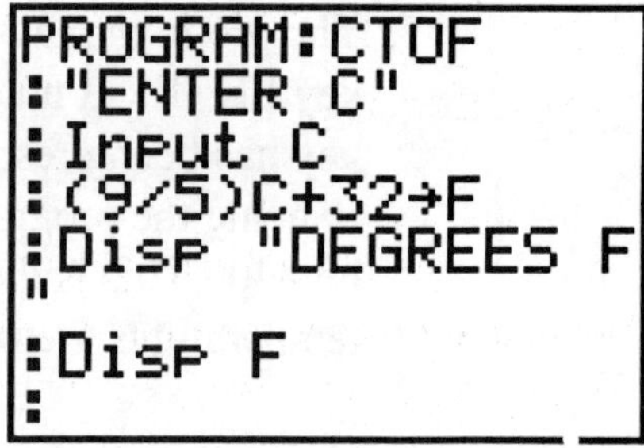

Figure 2.13

Exercises 2.5

Write programs to do the following.

1. Modify the equation $F = \frac{9}{5}C + 32$ to find values for C, given values for F. Write a program to perform this calculation. Evaluate for F = 212, 100, 98.6, 0, –40.
2. Write a program to evaluate $A = (1 + r)^5$ for values of r to be input. Evaluate for r = 0.05, 0.055, 0.06, 0.065
3. Write a program to evaluate the function $f(x) = x^4 - x^3 + x - 2$ for values of x. Evaluate for x = 0, 1, 2.5, 3.14, –2.34.
4. Write a program to evaluate the function $g(x) = x^5 - x^3 + 2$for values of x. Evaluate for x = 1, 2.5, –3.1, 1.998.
5. Write a program to evaluate the volume of a cylindrical tank for various values of r and h. The formula is $V = \pi r^2 h$. Evaluate for r = 123.4 and h = 25.6 and for r = 127.4 and h = 22.5.

6. Write a program to calculate miles per gallon. The number of miles driven and the number of gallons of gas required should be used as input. Evaluate for the case where it took 15.3 gallons to drive 345 miles.
7. Write a program to calculate the interest earned on P dollars deposited for t years at an annual interest rate of r percent per year (entered as a decimal) compounded n times per year. The formula is $P\left(1+\frac{r}{n}\right)^{nt}-P$.
   Find the interest earned on $1500.00 deposited for 5 years at an interest rate of 4.5% compounded 4 times per year.
8. Write a program to calculate the monthly payment M on a loan of P dollars for t years at an interest rate of r percent per year (entered as a decimal). The formula is $M=\frac{P}{K}$ where $K=\frac{1-(1+i)^{-12t}}{i}$ and $i=\frac{r}{12}$.
   Find the monthly payment on a loan of $60000.00 for a period of 20 years at an interest rate of 9.0% per year, then, at an interest rate is 10.0%.
9. Write and enter a program to solve a quadratic equation with real solutions.
10. Write and enter a program which accepts the coordinates of two points and gives the distance between them.
11. Enter the following program to graph straight line:

```
Disp "M = "
Input M
Disp "B = "
Input B
"MX + B"→Y1
DispGraph
```

   DispGraph is selected from the I/O menu when in programming mode.
12. Write and enter a program for graphing a quadratic function $y = ax^2 + bx + c$.
13. Write and enter a program for graphing a quadratic function $y = a(x + h)^2 + k$.

Chapter 3

Trigonometric Calculations

3.1 Performing Trigonometric Calculations
(Washington, Sections 4-3, 4-4, and 4-5)

Calculations involving trigonometric functions are done on the graphing calculator in much the same as with standard scientific calculators except that expressions are generally entered as they appear. For example, to find the sine of an angle, press the key: SIN, then enter the angle, and press ENTER. Care must be taken to be sure that the calculator is in the correct mode (degrees or radians).

Procedure C14. To change from radian mode to degree mode and vice versa.

1. Press the key: MODE
2. Move the cursor to either Radian or Degree on the third line depending on which mode is desired. (The mode which is highlighted is the active mode.) Press ENTER.
3. Press CLEAR or QUIT to return to the calculation screen.

Example 3.1: One leg of a right triangle is 6.5 units long. The angle adjacent to this leg is 37.5 degrees. Find the hypotenuse and other leg of the triangle.

Solution:

1. Let x represent the length of the other leg and r represent the length of the hypotenuse.
2. To find the other leg we make use of right triangle trigonometry to obtain

$$\tan 37.5° = \frac{x}{6.5} \quad \text{or} \quad x = 6.5 \tan 37.5°$$

   Make sure your calculator is in degree mode, then enter 6.5 tan 37.5 and press ENTER. The result is 4.9876 which when rounded to the correct number of decimal places is 5.0.
3. To find the hypotenuse we make use of right triangle trigonometry to obtain

$$\cos 37.5° = \frac{6.5}{r} \quad \text{or} \quad r = \frac{6.5}{\cos 37.5°}$$

Enter 6.5/cos 37.5 into your calculator and press ENTER. The result is about 8.193 or 8.2 when rounded to the correct number of decimal places. (Refer to Chapter 1 for guidelines on rounding.)
Thus, the other leg is 5.0 units long and the hypotenuse is 8.2 units long.

When we know the sine, cosine, or tangent of the angle and want to find the angle we use of the inverse trigonometric functions. The inverse sine function is written $\sin^{-1} x$, the inverse cosine function $\cos^{-1} x$ and the inverse tangent function $\tan^{-1} x$. For positive values of the trigonometric functions and for positive angles less than 90 degrees, we define the inverse trigonometric functions as follows:

$$\theta = \sin^{-1} x \text{ if and only if } x = \sin\theta$$
$$\theta = \cos^{-1} x \text{ if and only if } x = \cos\theta$$
$$\theta = \tan^{-1} x \text{ if and only if } x = \tan\theta$$
$$\theta = \sec^{-1} x \text{ if and only if } x = \sec\theta$$
$$\theta = \csc^{-1} x \text{ if and only if } x = \csc\theta$$
$$\theta = \cot^{-1} x \text{ if and only if } x = \cot\theta$$

By this definition, $\sin^{-1} x$ equals an angle θ if and only if the sine of θ is x.. Thus, if we know that the sine of an angle θ is a number n ($\sin\theta = n$), then it follows that $\theta = \sin^{-1} n$. The other inverse trigonometric functions may be thought of in a similar manner.

The inverse functions may be found by pressing the appropriate keys on the calculator. If we know the sine of an angle and want to find the angle, we enter SIN^{-1} followed by the sine of the angle, and press ENTER.

Example 3.2: One leg of a right triangle is 102 feet and the hypotenuse is 152 feet. Find the angle opposite the given leg.

Solution:

1. Let θ represent the angle.
2. From the trigonometry of a right triangle we know that $\sin\theta = \frac{102}{152}$

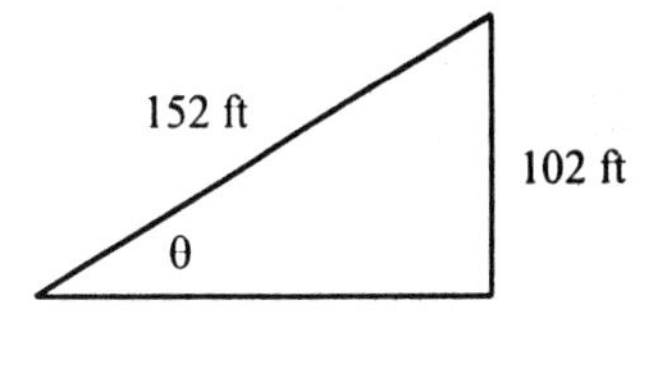

From the above definition this means that $\theta = \sin^{-1}\frac{102}{152}$.

3. Enter $\sin^{-1}(102/152)$ into your calculator and press ENTER. The result is about 42.148. Rounding this to the correct number of decimal places we find that $\theta = 42.1°$.

To find the values of the secant, cosecant or cotangent, we make use of the trigonometric identities:

$$\sec\theta = 1/\cos\theta \quad \text{or} \quad \sec\theta = (\cos\theta)^{-1}$$
$$\csc\theta = 1/\sin\theta \quad \text{or} \quad \csc\theta = (\sin\theta)^{-1}$$
$$\cot\theta = 1/\tan\theta \quad \text{or} \quad \cot\theta = (\tan\theta)^{-1}$$

For example, in order to find the secant of an angle, find the cosine of the angle then take the reciprocal. The reciprocal of a value can be found by dividing one by the value or by using the x^{-1} key on the calculator. Note that the –1 exponent has two different meanings. **When used with functions, the –1 exponent often means the inverse function, while when used in algebra or with a numerical value it means the reciprocal of the number.** The inverse sine of x is written as $\sin^{-1} x$ and the reciprocal of sin x is written as $(\sin x)^{-1}$.

Example 3.3: (a) Find csc 27.32° and (b) find θ if $\theta = \sec^{-1} 1.263$

Solution:

1. From the identities we know that $\csc 27.32° = \dfrac{1}{\sin 27.32°}$. We can use two methods of finding csc 27.32°.
   (a) Enter 1/sin 27.32 and press ENTER
   or
   (b) Enter (sin 27.32), then press the x^{-1} key.
   This appears on the screen as $(\sin 27.32)^{-1}$.
   Press ENTER.
   Why are the parentheses are necessary?
   The answer (properly rounded) is csc 27.32° = 2.179.
2. To find θ in part (b), we use the definition of the inverse trigonometric functions to note that if $\theta = \sec^{-1} 1.263$, then $\sec\theta = 1.263$.

   From the identities, this is the same as $\dfrac{1}{\cos\theta} = 1.263$.

   Solving this last equation for $\cos\theta$ we obtain $\cos\theta = \dfrac{1}{1.263}$.

Again using the definition of inverse functions we see that

$$\theta = \cos^{-1}\frac{1}{1.263} \quad \text{or} \quad \theta = \cos^{-1} 1.263^{-1}.$$

To find θ, enter either:

$\cos^{-1}(1/1.263)$ or $\cos^{-1}$ 1.263 followed by x^{-1}

and press ENTER. The result is $\theta = 37.65°$.

Exercise 3.1

In exercises 1-24, evaluate the indicated quantity. All angles are in degrees.

1. sin 23.5°
2. cos 45.7°
3. tan 7.5°
4. sin 0.01245°
5. sec 36.8°
6. cot 85.6°
7. cos 23.7°
8. tan 88.95°
9. csc 1.234°
10. sec 34.27°
10. cot 16.475°
12. csc 16.989°
13. $\sin^{-1}$ 0.8475
14. $\cos^{-1}$ 0.5245
15. $\cos^{-1}$ 0.0251
16. $\tan^{-1}$ 1.257
17. $\sec^{-1}$ 1.876
18. $\csc^{-1}$ 2.873
19. $\cos^{-1}$ 1.256
20. $\cot^{-1}$ 0.5621
21. $\cot^{-1}$ 1.54
22. $\sec^{-1}$ 1.05672
23. $\tan^{-1}$ 0.12376
24. $\sin^{-1}$ 1.1295
25. Find the sin A and cos B for each of the following:
    (a) A = 23.5° and B = 66.5°
    (b) A = 17.45° and B = 72.55°
    (c) A = 77.8° and B = 12.2°
    In each case, note that B = 90° – A. What conclusion can we arrive at about the sine and cosine of angles?
26. Find the tan A and cot B for each of the following:
    (a) A = 45.5° and B = 44.5°
    (b) A = 26.54° and B = 63.46°
    (c) A = 87.2° and B = 2.8°
    In each case, note that B = 90° – A. What conclusion can we arrive at about the tangent and cotangent of angles?
27. Evaluate $(\sin A)^2 + (\cos A)^2$ of each of the following:
    (a) A = 37.9°
    (b) A = 2.23°
    (c) A = 88.52°

What conclusion can we arrive at about this expression for any given angle A? Why is it necessary to enter $(\sin A)^2$ and $(\cos A)^2$ rather than $\sin A^2$ and $\cos A^2$?

28. Evaluate $(\sec A)^2 - (\tan A)^2$ for each of the following:
   (a) $A = 52.6°$
   (b) $A = 1.4°$
   (c) $A = 86.23°$
   What conclusion can we arrive at about this expression for any given angle A?
29. Evaluate $(\csc A)^2 - (\cot A)^2$ for each of the following:
   (a) $A = 33.3°$
   (b) $A = 0.565°$
   (c) $A = 83.28°$
   What conclusion can we arrive at about this expression for any given angle A?
30. Evaluate $\sin^{-1} x + \cos^{-1} x$ for each of the following:
   (a) $x = .25$
   (b) $x = .7895$
   (c) $x = 0.0125$
   What conclusion can we arrive at about this expression for any given value x?

Exercises 31-44 contain equations than may be encountered in solving problems in trigonometry. Evaluate the unknown in each equation.

31. $\sin x = \dfrac{2.456}{4.231}$
32. $\cos x = \dfrac{123.45}{63.88}$
33. $\tan x = \dfrac{0.612}{0.928}$
34. $\cot x = \dfrac{1.265}{8.231}$
35. $\sin 34.6° = \dfrac{x}{2.345}$
36. $\tan 45.65° = \dfrac{x}{16.24}$
37. $\cos 1.278° = \dfrac{24.35}{x}$
38. $\cot 23.7° = \dfrac{23.64}{x}$
39. $x^2 = 1.245^2 + 3.125^2 - 2(1.245)(3.125)(\cos 25.4°)$
40. $y^2 = 234.5^2 + 125.6^2 - 2(234.5)(125.6)(\cos 63.50°)$
41. $\sin B = \dfrac{25.2 \sin 23.2°}{16.1}$
42. $\sin C = \dfrac{6.43 \sin 65.5°}{7.25}$

43. $\cos A = \dfrac{16.2^2 + 12.5^2 - 17.3^2}{2(16.2)(12.5)}$

44. $\cos A = \dfrac{7.23^2 + 8.15^2 - 6.24^2}{2(7.23)(8.15)}$

45. In any triangle if b and c represent the lengths of two sides and θ the angle between these two sides, the area of the triangle can be found from the equation $A = \frac{1}{2} bc \sin \theta$. Use this formula to find the area of a triangle that has two sides of length 25.6 cm and 34.7 cm if the angle between these two sides is 56.52°.

♦ 46. *(Washington, Exercises 4-3, #43)* The voltage e at any instant in a coil of wire that is turning in a magnetic field is given by $e = E \cos \alpha$, where E is the maximum voltage and α is the angle the coil makes with the field. Find the acute angle α for E = 117 volts and
e = 25.5 volts, e = 37.0 volts, and e = 64.4 volts.

In exercises 47 - 54 solve the right triangle ACB for the missing parts. Angle C is the right angle.

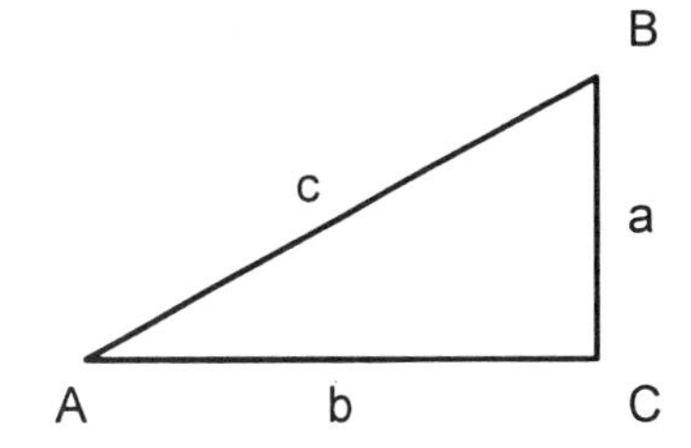

47. A = 23.41°, a = 25.25
48. B = 45.67°, a = 14.76
49. B = 45.8°. c = 2.25
50. A = 12.5°, c = 123
51. a = 163.5, c = 245.7
52. b = 45.67, c = 56.78
53. a = 0.1537, b = 0.9823
54. b = 15400, a = 25930

Hint for problems 55-58: Set up a formula and then evaluate the formula. Using a list of numbers may be helpful.

55. The center of a car's headlight is 25.0 inches above the surface of a highway. Determine the angle of depression needed for the center of the headlight beam to hit the highway 150.0, 200.0, 250.0, 300.0, and 350.0 feet in front of the car.
56. A person 72.0 inches high stands outside in the sunlight every even numbered hour from 6:00 A.M. till Noon. Determine the length of the shadow of this person at each even hour. Assume the sun rises at 6:00 A.M. and is directly overhead at noon and that the angle of elevation of the sun is directly proportional to the hour after sunrise.

♦ 57. *(Washington, Exercises 4-5, #9)* In designing a new building, the base of a doorway is 2.65 feet above the ground. A ramp for the disabled is to be built from to the base of the doorway along level ground. Determine the length along the ramp if the ramp makes angles with the ground of 4.0°, 4.5°, 5.0°, 5.5°, and 6.0°

♦ 58. *(Washington, Exercises 4-5, #12)* A guardrail is to be constructed around the top of a circular observation tower. The diameter of the observation area is 10.00 m. The guardrail is to be constructed in equal straight sections. Find the length of each section and also the total length of all sections if 5, 10, 15, 20, 25, 30, 35, or 40 sections are used. What would be the maximum total length of guardrail needed?

59. Write and test a program to solve a right triangle if:
    (a) Two legs are given.
    (b) One leg and the hypotenuse is given.
    (c) One angle and an adjacent leg is given.
    (d) One angle and an opposite leg is given.
    (e) One angle and the hypotenuse is given.

Chapter 4

Systems of Linear Equations

4.1 Graphical Solutions of Systems of Linear Equations
(Washington, Section 5-3)

A linear equation is an equation that may be written in the form $ax + by = c$ where a, b, and c are real numbers. A graph of such an equation is a straight line. A system of two such equations has a solution if there exists a pair of numbers (for x and y) which satisfy both equations. In this section we will see how a solution may be found by graphical means. Some applications problems that may be difficult to solve by algebraic means are readily solved by graphical means.

Example 4.1: Consider the system of equations:

$$3x + 2y = 6$$
$$x - 3y = 13$$

Determine 3 points on the graph of the first equation and 3 points on the graph of the second equation. By trial and error find a point which is on the graph of both equations.

Solution:

1. For the first equation, if we let x be 0, 2, and −2 and solve for y, we obtain pairs of numbers that satisfy the first equation and thus give three points on its graph: (0, 3), (2, 0), and (−2, 6).
2. Likewise in the second equation if we let x be 0, 2, and −2, we obtain the pairs of numbers (0, 13), (2, −11/3), and (−2, −5) that satisfy the second equation and give three points on its graph.
3. If the search for points is continued through enough x values, eventually (perhaps an eternity) it would be found that a pair of numbers (4, −3) satisfy both equations. This would correspond to the point where the two graphs intersect. There is only one such point that will satisfy both these equations since there is exactly one point of intersection of the graphs of the two lines if they are not parallel and do not coincide.

Because of the nature of the equations in Example 4.1, it was relatively easy to find a pair of numbers that satisfied both equations. For most systems of equations such a guessing game would be a ridiculous way of determining the desired values.

In order to solve a system of equations, we desire to find the set of numbers that satisfy all the equations if such numbers exist. A system of two linear equations in two unknowns may have no solutions, one solution, or an infinite number of solutions. A system of equations that has no solutions is called **inconsistent** and a system of equations that has an infinite number of solutions is called **dependent**.

To find the solution of a system of two linear equations that has one solution, we need to graph both equations on the same graph and find the point of intersection. To graph a linear equation, we first solve the equation for the dependent variable (y) and then follow Procedure G1.

Procedure G15. Finding an intersection point of two graphs.

Enter the functions into the calculator and adjust the viewing rectangle so the region of the graph containing the intersection point on the screen. Then, either

Zoom in on the point of intersection. To obtain the desired degree of accuracy, use the trace function to move the cursor just to the left and then just to the right of the intersection point after each zoom. When the
x-values, on either side of the intersection point, rounded off to the desired number of significant digits are equal, we have reached a solution. (The y-values may also be checked by moving the cursor just above and just below the intersection point.)

or

Determine the point of intersection by the built-in procedure:

1. Select the correct mode to find the point of intersection.

On the TI-82:	On the TI-85:
Press the key CALC Select 5:intersect	With the graph menu on the screen, select MATH (after pressing MORE). Then, select ISECT (after pressing MORE).

2. Move the cursor near the point of intersection.
3. If the cursor is on one of the graphs, continue. Otherwise, use the up and down cursor keys to move the cursor to the correct graph. (On the TI-82 the words "First curve?" appear on the screen.)
4. Press ENTER.
5. If the cursor is on the correct second graph, continue. Otherwise, use the up and down cursor keys to change the cursor to the correct graph. (On the TI-82 the words "Second curve?" appear.)

6. Press ENTER.
7. Obtain the coordinates of the intersection point.

   On the TI-82:
   The word "Guess?" appears on the screen. Make sure the cursor is near the point of intersection and press ENTER.
   The word Intersection and the coordinates of the intersection point will be given at the bottom of the screen.

   On the TI-85:
   The word ISECT and the coordinates of the intersection point will be given at the bottom of the screen.

Example 4.2: Solve the following system of equations. Find the x- and y-values to three decimal places.

$$5x + 2y = 7$$
$$3x = y + 3$$

Solution:

1. First solve each equation for y as a function of x.

$5x + 2y = 7$	and	$3x = y + 3$
$2y = -5x + 7$		$3x - 3 = y$
$y = (-5/2)x + 7/2$		$y = 3x - 3$

2. On the standard viewing rectangle, use Procedure G1 to graph $y = (-5/2)x + 7/2$ and $y = 3x - 3$ on the same graph.
3. Use the trace function (Procedure G4) to move the cursor near the point of intersection. For one value of x the graph will be on one side of the intersection point and for another value, on the other side. The intersection point must be between these two points. In this case for approximately x = 1.1579 the cursor is to the left of the intersection point and for x = 1.3684 it is on the right hand side. Leave the cursor on the point where x = 1.3684. (Your values may differ slightly.)

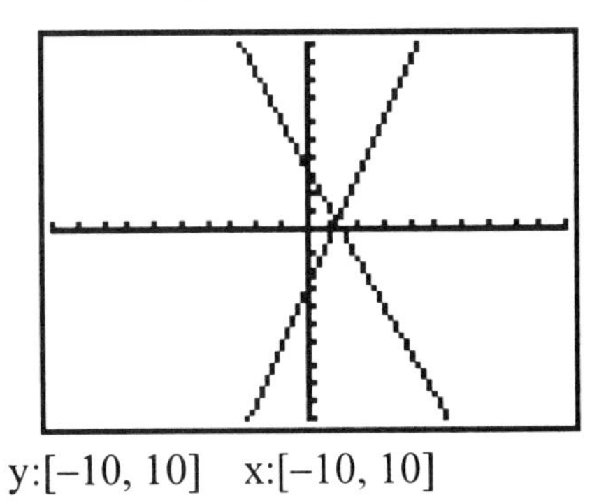

y:[−10, 10] x:[−10, 10]

Figure 4.1

4. Use Procedure G15 to find the point of intersection. The result, to a precision of three decimal places, is x = 1.182, y = 0.545.

It may happen that when trying to find the solution to a system of equations on a graphing calculator, the intersection point falls outside the viewing rectangle. In this case it is necessary to change the viewing rectangle so that the intersection point falls on the screen. Three different methods may be used to do this:

1. Change the viewing rectangle as described in Procedure G5. This may be particularly useful if the x or y values are very large or very small.
2. Use zoom out as described in Procedure G9.

or

3. Use the trace function to move the cursor near the point of intersection until the point of intersection lies on the screen. This works only if the point of intersection lies off the right or off the left of the screen. In this case the screen will scroll (move over) in the direction the cursor is moving.

Example 4.3: Solve the following system. Find answer to two decimal places.

$$x + 10y = 50$$
$$x - 2y = 8$$

Solution: 1. Solve each equation for y as a function of x.

$$x + 10y = 50 \quad \text{and} \quad x - 2y = 8$$
$$10y = -x + 50 \qquad -2y = -x + 8$$
$$Y = \frac{-1}{10}x + 5 \qquad y = \frac{1}{2}x - 4$$

2. Graph y = (–1/10)x + 5 and y = (1/2)x – 4 on the same graph using the standard viewing rectangle. The graph is in Figure 4.2.
3. From the graph it does appear that the lines do intersect, but off the screen to the right. In this case we use the trace function and move the cursor along a graph until the point of intersection is on the screen.

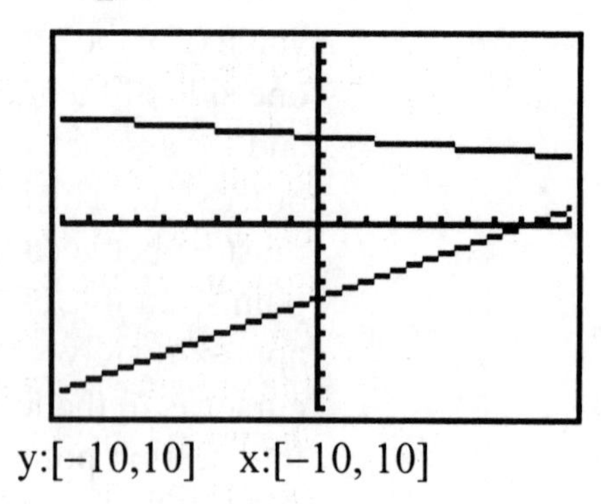

y:[–10,10] x:[–10, 10]

Figure 4.2

4. Use Procedure G15 to find the point of intersection. The solution is $x = 15.00$ and $y = 3.50$.

Sometimes the values of the variables in the equations are such that when the standard viewing rectangle is used either the point of intersection is not apparent or one or both graphs do not appear on the screen. In such a situation a better viewing rectangle must be used based on estimates of the values of the variables. The viewing rectangle may then be adjusted as seem appropriate. One method of estimating the maximum or minimum y-values that are needed is to use the trace function.

♦ Example 4.4: *(Washington, Section 5-3, Example 4)* A driver traveled for 1.5 hours at a constant speed along a highway. Then, through a construction zone, the driver reduced the car's speed by 20.0 miles per hour for 30.0 minutes. If 100.0 miles were covered in the 2.0 hours, what were the two speeds.

Solution:

1. We recall that distance equals rate times time and let x = the car's highway speed and y = the car's speed in the construction zone. Then, 1.5x represents the distance traveled on the highway and 0.50x represents the distance traveled in the construction zone and the sum gives the total number of miles traveled.

$$1.5x + 0.50y = 100.$$

Since the difference in speeds is 20 miles per hour

$$y = x - 20$$

2. Solve each equation for y, enter into the calculator and graph on the standard viewing rectangle. In this case, nothing appears on the screen since both graphs are located outside the viewing rectangle. Using the trace function and observing the x- and y-intercepts of the functions, we determine that a reasonable viewing rectangle is

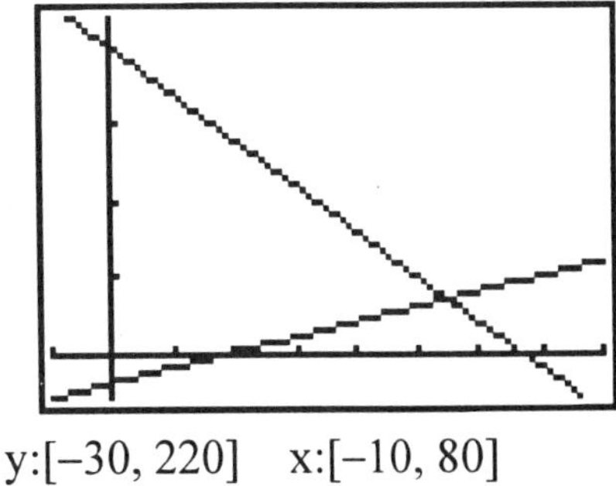

y:[−30, 220] x:[−10, 80]

Figure 4.3

Xmin = – 10, Xmax = 80
Ymin = –30, Ymax = 220
This gives the graph in Figure 4.3.

3. Use Procedure G15 to find the point of intersection. Based on the accuracy of the values given in the problem, we need to round off the answers to two significant digits. The answer is $x = 55$, $y = 35$. Thus, the highway speed is 55 miles per hour and the construction zone speed is 35 miles per hour.

When we solve a system of two linear equations by graphing and parallel lines are obtained, there are no points of intersection and thus no solutions and the system is an inconsistent system. The lines can be verified as being parallel by using the trace function. To do this move the cursor to a point on one graph, jump the cursor to the corresponding point on the second graph at the same x-value. Note the difference between the y-values. Repeat this process for several points. If the differences of the y-values is always the same, the lines are parallel.

In a dependent system of two equations, only one line will appear on the graph and each point on the line satisfies both equations. In this case there are an infinite number of solutions.

The method of finding the intersection of two graphs may also be used to determine an x-value that gives a particular y-value. To do this, graph the given function and also the function $y = k$ where k is the y-value. Then, find the intersection point.

Exercise 4.1

Solve each of the following systems of linear equations. Find each answer to 3 significant digits. Identify any systems that are inconsistent or dependent.

1. $2x - 3y = 5$
   $4x + 2y = 3$

2. $3x + 4y = 4$
   $2x - 5y = 7$

3. $2x + 3y = 35$
   $4y = 3x - 20$

4. $1.3x = 5.4y + 6.3$
   $2.6y + 0.60x + 19.7 = 0$

5. $2.40x + 1.30y = 7.80$
   $1.68x + 0.91y = 5.46$

6. $11x + 25y = 95$
   $2x = 3y - 49$

7. $5.24x - 3.18y = 25.67$
   $3.72x = 4.29y - 37.35$

8. $7.6x - 4.4y = 6.30$
   $5.7x = 3.3y + 8.63$

9. $127x + 245y = 670$
   $25.4x + 49.0y = 402$

10. $60.6x + 50.2y = 233.7$
    $23.7y = 27.2 - 24.8x$

11. A company buys two types of disks for personal computers on two different orders and they do not know the cost of each disk of each type. They do know that in one order they received 425 of type A and 575 of type B and the cost was $675. In the second order they received 750 of type A and 225 of type B and the cost was $585. Determine the cost of each type A and each type B disk.

♦ 12. *(Washington, Exercises 5-3, # 32)* A total of 42.5 tons of two types of ore is to be loaded into a smelter. The first type contains 6.25% copper and the second contains 2.35% copper. Find the necessary amounts of each ore (to three significant figures) to produce 2.50 tons of copper.

♦ 13. *(Washington, Exercises 5-4, # 35)* Two grades of gasoline are mixed to make a gasohol blend. The first grade contains 4.650% alcohol and the second grade contains 12.25% alcohol. It is desired to mix x liters of the first grade with y liters of the second to obtain 12,225 liters of a mixture containing 9.500% alcohol. The equations are:

$$0.04650x + .1225y = .09500(12225)$$
$$x + y = 12225$$

Find how many liters of each mixture are needed.

14. Given the two equations

$$-11.5x + 13.8y = 18.4$$
$$22.08y = 18.4x + 29.44$$

Graph these equations. What can you tell about the solution? Which, if any, of the following are solutions to this system?

$x = 20, y = 18$; $x = 32, y = 28$; $x = 56, y = 48$; $x = 68, y = 58$; $x = 86, y = 73$.

15. A cost function for a business is given by $y = 1225 + 22.5x$. The income is given by the revenue function $y = 35.75x$. In both cases, y is in dollars and x is the number of items. Graph these functions on the same graph then answer these questions.
    (a) Determine the cost and revenue when $x = 46$ items by finding the difference in y-values. (Hint: Jump the cursor between the curves.) If the business produces and sells 46 items is the business making any profit?
    (b) Determine the cost and revenue when $x = 138$ items by finding the difference in y-values. If the business produces and sells 138 items is it making any profit? If so, how much?

(c) Determine the point at which the cost equals the revenue. This is called the break even point. What is the profit at this point?

16. The cost function for a business is given by $y = 18725 + 125x$ and the income is given by the revenue function $y = 175x$ where y is in dollars and x is the number of items. Graph these functions on the same graph then answer these questions.
    (a) Determine the cost and revenue when $x = 200$ items by finding the difference in y-values. If the business produces and sells 200 items is the business making any profit?
    (b) Determine the cost and revenue when $x = 500$ items by finding the difference in y-values. If the business produces and sells 500 items is it making any profit? If so, how much?
    (c) Determine the point at which the cost equals the revenue. This is called the break even point. What is the profit at this point?

In certain cases the calculator may be used to solve systems of three equations. One case in which this is true is when one of the equations contains only two variables and the missing variable may be easily eliminated from the other equations.

◆ 17. *(Washington, Exercises 5-6, # 19)* To solve the system of equations:

$$\begin{aligned} 0.707F_1 - 0.800F_2 \qquad\quad &= 0 \\ 0.707F_1 + 0.600F_2 - F_3 &= 10.0 \\ 3.00F_2 - 3.00F_3 &= 20.0 \end{aligned}$$

Subtract the first equation from the second to obtain the set of equations:

$$\begin{aligned} 1.400F_2 - F_3 &= 10.0 \\ 3.00F_2 - 3.00F_3 &= 20.0 \end{aligned}$$

Solve these equations on the graphing calculator. Once values for F_2 and F_3 have been found, F_1 may be readily found from the first equation.

◆ 18. *(Washington, Exercises 5-6, # 20)* In applying Kirchhoff's laws to an electric circuit, a student comes up with the following equations for currents I_A, I_B, and I_C (in amperes):

$$\begin{aligned} I_A + I_B + I_C &= 0 \\ 3.5I_A - 8.8I_B \qquad &= 4.77 \\ -8.8I_B + 4.8I_C &= 5.58 \end{aligned}$$

Use the method described in Exercise 17 to solve this system of equations.

Chapter 5

Graphs of Trigonometric Functions

5.1 Graphing of Basic Sine and Cosine Functions
(Washington, Sections 10-1 and 10-2)

The graphing of trigonometric functions on the calculator is handled in much the same way as the graphing of algebraic functions. However, we need to be aware of two things. First, trigonometric functions are graphed with the value of the independent variable in radians. (It is important to be sure the calculator is in radian mode. See Procedure C6.) Second, there is a special trigonometric viewing rectangle for trigonometric functions (see Procedure G6) that marks the x-axis in terms of π.

Example 5.1: On the graphing calculator graph $y = \sin x$ and $y = \cos x$ using the trigonometric viewing rectangle. Determine the local maximum and local minimum points and estimate the x-intercepts to one decimal place.

Solution:

1. Graph $y = \sin x$ on the standard trigonometric viewing rectangle. (See procedure G6). For the standard trigonometric viewing rectangle:

   Xmin = –6.28... (2π), Xmax = 6.28... (2π),
   Xscl = 1.57... ($\pi/2$),
   Ymin = –3, Ymax = 3,
   Yscl = .25

2. The graph of the sin x appears in Figure 5.1.
3. The local maximum points are (–4.7, 1) and (1.5, 1) and the local minimum points are (–1.5, –1) and (4.7, –1).

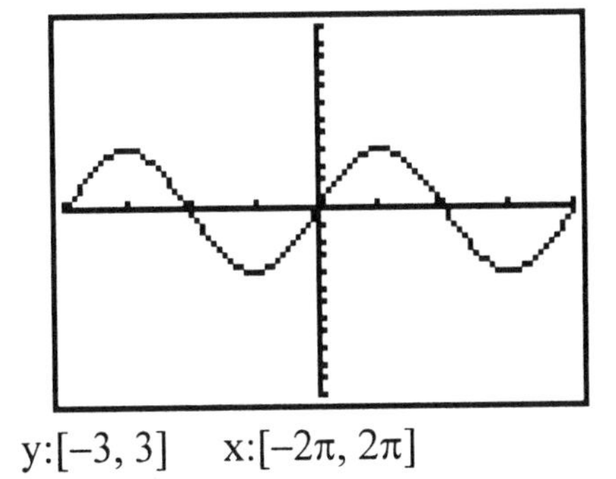

y:[–3, 3] x:[-2π, 2π]

Figure 5.1

These correspond to $(-\frac{3\pi}{2}, 1)$, $(\frac{\pi}{2}, 1)$, $(-\frac{\pi}{2}, -1)$ and $(-\frac{3\pi}{2}, -1)$.
The x-intercepts are –6.3, –3.1, 0, 3.1, 6.3 that correspond to $-2, \pi, -\pi, 0, \pi, 2\pi$.

4. Graph cos x on the standard trigonometric viewing rectangle. The graph is in Figure 5.2
5. The local maximum points are (–6.3, 1), (0, 1) and (6.3, 1) and the local minimum points are (–3.1, –1) and (3.1, –1). These correspond to $(-2\pi, 1)$, $(0, 1)$, $(2\pi, 1)$, $(-\pi, -1)$, and $(\pi, -1)$. The x-intercepts are –4.7, –1.5, 1.5, 4.7 which correspond to $-\frac{3\pi}{2}, -\frac{\pi}{2}, \frac{\pi}{2}$, and $\frac{3\pi}{2}$.

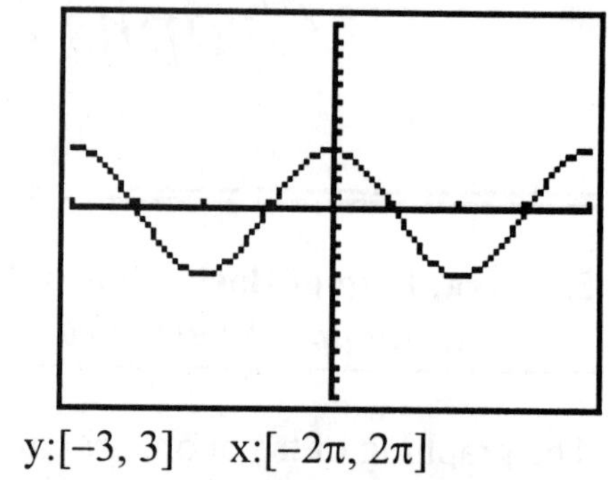

y:[–3, 3] x:[-2π, 2π]

Figure 5.2

A mental snapshot should be taken of the graphs of sin x and cos x (commit to memory). It is important to know what these graphs look like, the location of their local maximum and local minimum points (in terms of π), and the location of the x-intercepts (in terms of π). Graphs of other sine and cosine functions may be obtained from these basic graphs by modifying the graph. The exercises for this section will explore this more. The graphs shown in Example 5.1 are typical of graphs of trigonometric functions in that a portion of the curve repeat along the x-axis. The repeating portion of the curve is called a **cycle.**

The **amplitude** of the graph of a the sine or cosine is one-half the difference between the minimum and the maximum y-values. The **period** is the length of one complete cycle of the graph. If P is the period of a function, then the value of the function evaluated at x and at x + P is the same. That is, $f(x) = f(x + P)$. The amplitude for both y = sin x and y = cos x of Example 5.1 is 1 and the period in each case is 2π.

Example 5.2: Graph y = 2 cos 0.5x and y = 3 cos 0.5x and determine the difference between the two graphs.

Solution:

1. Make sure the calculator is in radian mode, then graph y = 2 cos (0.5x) and y = 3 cos (0.5x) on the trigonometric viewing rectangle. Although the parentheses are not needed, it is always a good idea to include the quantity following a trigonometric function within parentheses.

2. The resulting graph is in Figure 5.3. Note that the only difference in the two graphs are the heights or amplitudes of the curves. The amplitude of the first graph is 2 and of the second graph is 3. Both curves cross the x-axis at the same points. The graph of $y = 3 \cos(0.5x)$ may be obtained from the graph of $y = \cos(0.5x)$ by stretching the graph in the y-direction.

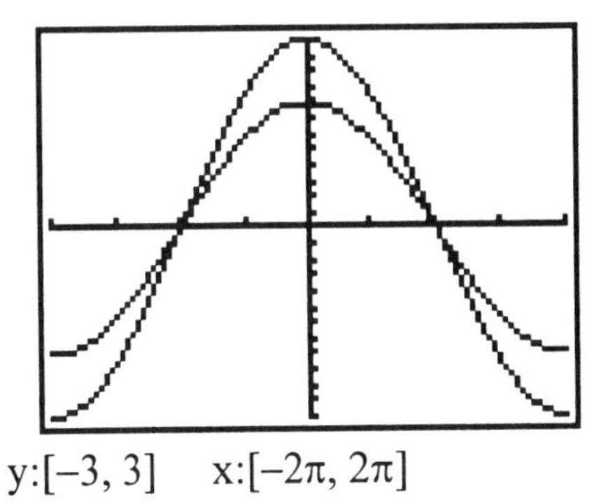

y:[−3, 3] x:[−2π, 2π]

Figure 5.3

The location of the x-intercepts and of the highest or lowest points on the graphs may be determined by using the trace key as before. In order to determine these points to more precision, we may use the zoom function or other functions of the calculator.

The cosine graph of Example 5.1 was stretched in the y-direction by multiplying the function by a constant greater than one. If we multiply by a constant less than one, the graph will be compressed in the y-direction. The graph may also be stretched or compressed in the x-direction or reflected about the x-axis. These properties will be further investigated in the exercise set.

Care is needed in determining the period of a function from the graph on the calculator screen. Because of the way the calculator plots the graph, things may not be what they appear. (See Exercises 7 and 8 in Exercise Set 5.1) Before graphing trigonometric functions on the graphing calculator it is best to first determine the amplitude and period. This will allow the correct choice of a viewing rectangle.

Exercise 5.1

For each of the following use the trigonometric viewing rectangle. Make sure calculator is in radians. Find values to two decimal places.

1. Use the calculator to graph $y = \sin x$, $y = 2 \sin x$, and $y = 3 \sin x$.
   (a) What are the y-coordinates of the local maximum and local minimum points in each case?
   (b) What is the amplitude of each function?
   (c) What is the effect of changing A in the equation $y = A \sin x$?
   (d) Sketch the graph of $y = 5 \sin x$ by hand. Check your answer on the calculator.

2. Use the calculator to graph $y = \cos x$, $y = 3 \cos x$, and $y = 6 \cos x$.
   (a) What are the y-coordinates of the local maximum and local minimum points in each case?
   (b) What is the amplitude of each function?
   (c) What is the effect of changing A in the equation $y = A \cos x$?
   (d) Sketch the graph of $y = 4 \cos x$ by hand. Check your answer on the calculator.
3. Use the calculator to graph $y = 2 \sin x$, $y = 3 \cos x$, $y = -2 \sin x$ and $y = -3 \cos x$.
   (a) What are the y-coordinates of the local maximum and local minimum points in each case?
   (b) What is the amplitude of each function?
   (c) What is the effect of changing the sign of A in the equation $y = A \sin x$ or in the equation $y = A \cos x$?
   (d) Sketch the graph of $y = -4 \sin x$ and $y = -6 \cos x$ by hand. Check your answers on the calculator.
4. Use the calculator to graph $y = 2 \sin x$, $y = 2 \sin 2x$, and $y = 2 \sin 4x$.
   (a) In each case, what are the first three x-intercepts that have a value greater than or equal to zero? Determine the difference between the first and third. What is this in terms of π?
   (b) What is the amplitude and period of each function?
   (c) What is the effect of changing B in the equation $y = A \sin Bx$?
   (d) Sketch the graph of $y = 5 \sin 3x$ by hand. Check your answer on the calculator.
   (e) Sketch the graph of $y = -3 \sin 2x$ by hand. Check your answer on the calculator.
5. Use the calculator to graph $y = 4 \cos x$, $y = 4 \cos 2x$, and $y = 2 \cos 3x$.
   (a) In each case, what are the first three x-intercepts that have a value greater than or equal to zero? Determine the difference between the first and third. What is this in terms of π?
   (b) What is the amplitude and period of each function?
   (c) What is the effect of changing B in the equation $y = A \cos Bx$?
   (d) Sketch the graph of $y = 5 \cos 2x$ by hand. Check your answer on the calculator.
   (e) Sketch the graph of $y = -6 \cos 5x$ by hand. Check your answer on the calculator.
6. Determine the amplitude and period of each of the following functions, then sketch one cycle of the graphs by hand. Indicate on the x-axes the intercepts and on the y-axis the amplitude. You may use your calculator to check your answer.
   (a) $y = 2.5 \sin 3x$ (b) $y = 15.0 \cos 2x$

(c) $y = 6 \sin \pi x$ (d) $y = -5 \cos (\frac{\pi}{3}x)$

(e) $y = 110 \cos 60\pi x$ (f) $y = -0.5 \sin 2.5x$

7. For the function $y = 3 \sin 50x$:
   (a) Graph on the standard trigonometric viewing rectangle. What does the period appear to be from the graph?
   (b) What is the real period of this function?
   (c) Graph this function so that two cycles appear on the screen for positive x-values. What are the minimum and maximum values of x that describe the viewing rectangle?
8. For the function $y = 2 \cos 60\pi x$:
   (a) Graph on the standard trigonometric viewing rectangle. Estimate the period from this graph.
   (b) What is the real period of this function?
   (c) Graph this function so that two cycles appear on the screen for positive x-values. What are the minimum and maximum values of x that describe the viewing rectangle?

In exercises 9-12, graph the function so that two cycles appear on the screen for positive x-values. In each case, give the minimum and maximum values of x and y that describe the viewing rectangle and estimate the x-intercepts.

9. $y = 5 \sin 25x$
10. $y = 0.25 \cos 18x$
11. $y = 0.75 \cos 0.024\pi x$
12. $y = 50 \sin 0.10x$

13. Use the calculator to graph $y = 20 \sin 300x$ in the trigonometric viewing rectangle. Is the graph you obtain correct? Can you explain what happened? What would be a better viewing rectangle in which to view this graph? Find the smallest positive x-intercept to two significant digits.
14. Use the calculator to graph $y = 120 \cos 120\pi x$ in the trigonometric viewing rectangle. Is the graph you obtain correct? Can you explain what happened? What would be a better viewing rectangle in which to view this graph? Give the values of Xmin, Xmax, Ymin, and Ymax that you use. Find the smallest positive x-intercept to two significant digits.

◆ 15. *(Washington, Exercises 10-2, #45)* The standard electric voltage in a 60-Hz alternating current circuit is given by $V = 120 \sin 120\pi t$ where t is the time in seconds.
   (a) Graph and from the graph determine the values for V when $t = 0.008$ and when $t = 0.010$ seconds.
   (b) For what values of t ($0 \le t \le 0.02$) is the voltage zero?
   (c) Find the first and second positive times when the voltage is at its maximum.

◆ 16. *(Washington, Exercises 10-2, #46)* The end of a tuning fork moves with a displacement given by $y = 1.60 \cos 460\pi t$, where y is in millimeters and t is in seconds.

(a) Graph and from the graph determine the values of y when $t = 0.0025$ seconds and when $t = 0.010$ seconds.

(b) For what values of t $(0 \le t \le 0.005)$ is $y = 1.60$?

(c) Find y when $t = 0.003$ and when $t = 0.010$ seconds.

5.2 Graphing of Sine and Cosine Functions with Displacement *(Washington, Section 10-3)*

We now investigate the graphs of functions of the form $y = a \sin(bx + c)$ and $y = a \cos(bx + c)$. The value c is referred to as the **phase angle**. The affect of c is to shift the graph horizontally. The amount of shift is often referred to as the **displacement** or **phase shift**.

Example 5.3: Graph the functions $y = 3 \sin x$ and $y = 3 \sin(x + \frac{\pi}{4})$ on the same graph. Determine how far and in which direction the second graph is shifted relative to the first graph and express in terms of π.

Solution:

1. Enter the first function for Y_1 and the second function for Y_2. Be sure the calculator is in radian mode and use the trigonometric viewing rectangle. The result is shown in Figure 5.4.

2. To determine how far the second curve is shifted horizontally from the first, use the trace function to determine the first positive x-intercept of the first graph. This occurs at $x = 3.1415927$. Then use the trace function to determine the first positive x-intercept of the second graph. This is at $x = 2.3561945$. The difference between the two intercepts is 0.7853982 or $\frac{\pi}{4}$. We also note that the second curve is shifted to the left relative to the first curve.

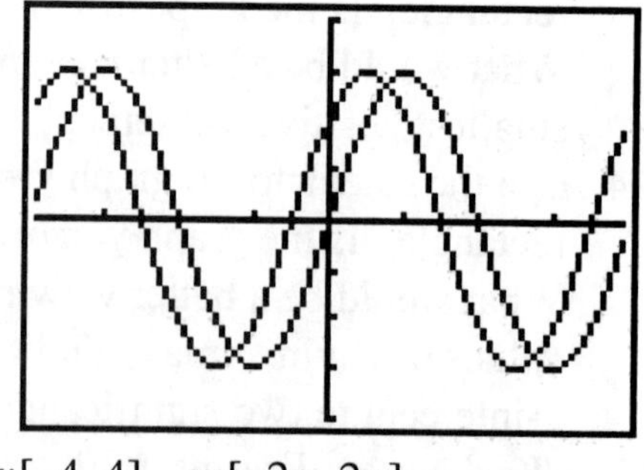

y:[−4, 4] x:[−2π, 2π]

Figure 5.4

We conclude that the addition of the phase angle $+\frac{\pi}{4}$ to the function y = sin x, displaces the curve $\frac{\pi}{4}$ units to the left. Another way of stating this is that replacing x by $x + \frac{\pi}{4}$ in the function y = sin x shifts the graph $\frac{\pi}{4}$ units to the left.

Exercise 5.2

Determine values to two decimal places.

1. Graph the functions $y = 2 \sin x$, $y = 2 \sin(x + \frac{\pi}{4})$, and $y = 2 \sin (x + \frac{\pi}{2})$.
   (a) What is the amplitude and period of each graph?
   (b) Determine the x-values where each graph crosses the x-axis for $0 < x \leq \pi$. How much is the second graph shifted with respect to the first? What about the third with respect to the first? In what direction? Express each difference in terms of π. (Example: $1.57 = \frac{\pi}{2}$.)
   (c) What is the effect of C in the equation $y = A \sin (x + C)$?
   (d) Does the amplitude have any effect on the displacement? If so, what?
   (e) Sketch one cycle of the graph of $y = 3 \sin (x + \frac{\pi}{3})$. Check your answer on the calculator.
2. Graph the functions $y = 3 \cos x$, $y = 3 \cos(x + \frac{\pi}{4})$, and $y = -2 \cos (x - \frac{\pi}{4})$.
   (a) What is the amplitude and period of each graph?
   (b) Determine the x-values where the graph crosses the x-axis for $0 < x \leq \pi$. How much is the second graph shifted with respect to the first? What about the third with respect to the first? In what direction? Express each difference in terms of π. (Example: $1.57 = \frac{\pi}{2}$.)
   (c) What is the effect of C in the equation $y = A \cos (x + C)$?
   (d) Does the amplitude have any effect on the displacement? If so, what?
   (e) Sketch one cycle of the graph of $y = -3 \cos (x - \frac{\pi}{2})$. Check you answer on the calculator.

3. Graph the functions $y = 2 \sin 2x$, $y = 2 \sin(2x - \frac{\pi}{2})$, and $y = 2 \sin(2x - \frac{\pi}{4})$.
   (a) What is the amplitude and period of each graph?
   (b) Determine the x-values where the graph crosses the x-axis for $0 \le x \le \pi$. How much is the second graph shifted with respect to the first? What about the third with respect to the first? In what direction? Express each difference in terms of π.
   (c) Divide the phase angle by the coefficient of x in each case and note how this relates to the displacement. What is the effect of $\frac{C}{B}$ in the equation $y = A \sin(Bx + C)$?
   (d) Sketch one cycle of the graph of $y = 3 \sin(2x + \frac{\pi}{4})$. Check your answer on the calculator.
4. Graph the functions $y = 2.5 \cos \pi x$, $y = 2.5 \cos(\pi x + \frac{\pi}{3})$, and $y = 2.5 \cos(\pi x - \frac{\pi}{6})$.
   (a) What is the amplitude and period of each graph?
   (b) Determine the x-values where the graph crosses the x-axis for $0 < x \le 1$. How much is the second graph shifted with respect to the first? What about the third with respect to the first? In what direction? Express each difference in terms of π.
   (c) Divide the phase angle by the coefficient of x in each case and note how this relates to the displacement. What is the effect of $\frac{C}{B}$ in the equation $y = A \sin(Bx + C)$?
   (d) Sketch one cycle of the graph of $y = 2 \cos(\pi x + \frac{\pi}{4})$. Check your answer on the calculator.
5. For each of the following equations, determine (1) the amplitude, (2) the period, and (3) the displacement. Then make a sketch of one cycle of each by hand. Check your answer with the calculator.
   (a) $y = 5 \cos 0.5x$
   (b) $y = 3 \sin(4\pi x - \pi)$
   (c) $y = 2.5 \cos(x - \frac{\pi}{4})$
   (d) $y = -3.5 \sin(2x + 0.5)$
6. For each of the following equations, determine (1) the amplitude, (2) the period, and (3) the displacement. Then make a sketch of one cycle of each by hand. Check your answer with the calculator.
   (a) $y = 12 \sin(0.25x + 0.25)$
   (b) $y = -6 \cos(3\pi x - 0.50)$
   (c) $y = -4.5 \cos(6x - 3\pi)$
   (d) $y = 24.5 \sin(6\pi - 3\pi)$

7. Use the graphing calculator to graph each of the following. In each case for $0 \leq x \leq \pi$, determine (1) the x-intercepts, and (2) the local maximum and local minimum points. Find all values to two decimal places.
   (a) $y = 2\sin(\pi x)$ (b) $y = 3.5\cos(2.4x + 1.63)$
8. Use the graphing calculator to graph each of the following. In each case for $0 \leq x \leq \pi$, determine (1) the x-intercepts, and (2) the local maximum and local minimum points. Find all values to two decimal places.
   (a) $y = 2\sin(\pi x)$ (b) $y = 4\sin(2x - \frac{\pi}{2})$
9. Graph one cycle of the function $y = 57.5\sin(0.50\pi t - 0.33)$. Then
   (a) Determine y when $x = 1.75$ and $x = 2.75$.
   (b) Determine x when $y = 50.0$ and $y = -45.5$.

◆ 10. *(Washington, Exercises 10-3, #26)* The electric current i, in microamperes, in a certain circuit is given by $i = 3.8\cos 2\pi(t + 0.20)$, where t is the time in seconds. Graph three cycles of this graph then determine:
   (a) The current at $t = 1.00$ seconds and at $t = 2.50$ seconds.
   (b) The time s in the first three cycles that the current is 2.50 microamperes.

◆ 11. *(Washington, Exercises 10-3, # 27)* A certain satellite circles the earth such that its distance y, in miles north or south (attitude is not considered) from the equator is given by $y = 4500\cos(0.025t - 0.25)$, where t is the time (in min) after launch. Consider a positive y-value as being north of the equator. Graph two cycles of this curve then
   (a) Determine how far north or south of the equator the satellite is after one hour, after three hours, and after 6.5 hours.
   (b) Determine all the times during the first two cycles that the satellite is 4000 miles above the equator.

12. For the satellite in Exercise 11, assume each cycle corresponds to one revolution of the satellite about the earth. Due to the revolving of the earth about its axis, the satellite does not appear above the same points of the earth on each revolution. To see how the orbit of the satellite might appear when superimposed on a map of the earth, we need to determine the displacement relative to each orbit. To do this, first determine the number of cycles per day (divide 24×60 by the period) of the satellite, then, divide the period by the number of cycles per day. This last value will give the time displacement, D, of each orbit. To see several orbits, graph, on the same set of axes, the function $y = 4500\cos(0.025(t + nD) - .25)$ for $n = 0, 1, 2$, and 3 (n is one less than the number of the orbit).
   (Hint: It is helpful to store the value for D and use a list for n.)
13. The voltage of an alternating electric current is given by $e = 110\sin(120\pi t + \frac{\pi}{6})$.
   (a) Find the values of t for which the voltage e is zero for $t < 1/60$.
   (b) What is the first time that the voltage reaches a high point?

(c) What is the second time the voltage reaches a high point? How much does this differ from the value of part (b)? Does this seem reasonable?
(d) What is t when e = 50 for t < 1/60?
(e) What is e when t = 0.010 sec.?

14. An example of damped simple harmonic motion is given by the equation $y = 5e^{-25t}\cos(50\pi t)$. (Here e is the mathematical constant 2.7128...)
(a) Find the positive values of t for which the graph crosses the horizontal axis the first and second time.
(b) Find the y and t values for which the graph reaches its high points the first and second times.

◆ 15. *(Washington, Exercises 10-5, #9)* For a point on a string, the displacement y is given by $y = A \sin 2\pi(\frac{t}{T} - \frac{x}{\lambda})$. Each point on the string moves with simple harmonic motion according to this equation. Graph two cycles for the case where A = 3.25 cm, T = 0.0445 s, λ = 42.5 cm and x = 5.45 cm. Then, determine
(a) The displacement after 0.0220 s and after 0.0500 s.
(b) The times t in the first two cycles when the displacement is –2.50 cm.

◆ 16. *(Washington, Exercises 10-5, #16)* The acoustical intensity of a sound wave is given by $I = A\cos(2\pi f t - \alpha)$, where *f* is the frequency of the sound. Graph two cycles of I as a function of t if $A = 0.0245\ \text{W/cm}^2$, $f = 245$ Hz, and $\alpha = 0.775$. Then, determine
(a) The acoustical intensity when t = 0.00100 and when t = 0.00600.
(b) The time t in the first two cycles when the acoustical intensity is a maximum.

17. In each of the following cases, graph y_1 and y_2 on the same set of axes and determine how the graphs of the two functions compare. Which of the functions are equivalent?
(a) $y_1 = 2\sin x$ and $y_2 = 2\sin(-x)$
(b) $y_1 = 3\cos x$ and $y_2 = 3\cos(-x)$
(c) $y_1 = -4\cos x$ and $y_2 = 4\cos(x + \pi)$
(d) $y_1 = 3\sin x$ and $y_2 = 3\cos(\frac{\pi}{2} - x)$
(e) $y_1 = 3\cos(x - \frac{\pi}{2})$ and $y_2 = 5\cos(x + \frac{3\pi}{2})$

5.3 Other Trigonometric Graphs
(Washington, Sections 10-4 and 10-6)

Investigation of the graphs of tangent, cotangent, secant, and cosecant functions are left to the exercise set. To graph the cotangent, secant, and cosecant, enter the functions into the calculator as the reciprocal of tangent, cosine, and sine functions.

In some cases it is helpful to see the graphs of two separate functions, then the graph of the sum of these functions.

Procedure G16. To graph each of two functions and then their sum.

1. Set the correct values for the viewing rectangle.
2. Enter the first function into the function table for Y_1.
3. Enter the second function into the function table for Y_2.
4. Move the cursor after the equal sign for Y_3. Press the Y-VARS key and select Y_1 from the menu. (See Procedure C11)
5. Press the key: +
6. Press the Y-VARS key and select Y_2 from the menu. (The third line should now appear as $:Y_3=Y_1+Y_2$)
7. Press GRAPH

The graphing of composite trigonometric functions is illustrated in Example 5.4. When graphing composite trigonometric functions, the viewing rectangle should be set such that at least one complete cycle of the composite function is graphed. The period of the composite function should be taken as the least common multiple of the period of each of the separate functions and the amplitude of the composite function may have an amplitude equal to the sum of the amplitudes of the separate functions.

Example 5.4: Graph the function y = 2 sin x + 0.5 cos 6x.

Solution:

1. Select an appropriate scale for the x and y variables. The amplitude of the composite functions may be 2 + 0.5 or 2.5. The period of the first function is 2π and or the second function is $\frac{\pi}{3}$. For the composite function we take 2π which is the least common multiple of these two numbers. In order to see at least one complete cycle the x-scale needs to be at least 2π in length. Set Xmin = 0, Xmax = 6.28 (2π), Xscl = 1.57 ($\frac{\pi}{2}$), Ymin = −3, Ymax = 3, and Yscl = 1.

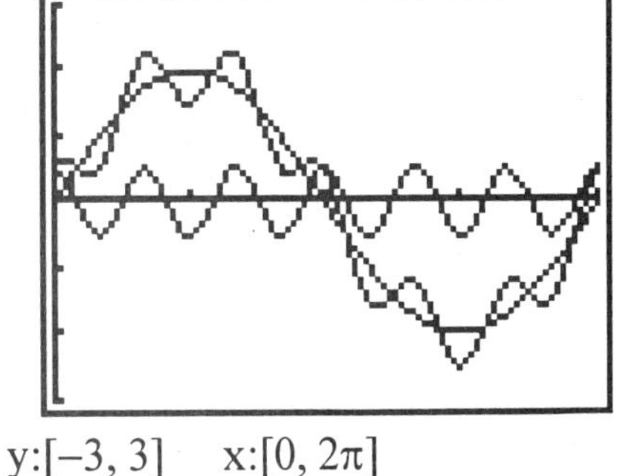

y:[−3, 3] x:[0, 2π]

Figure 5.5

2. The function y is the sum of the two functions Y_1 = 2 sin x and Y_2 = 0.5 cos 6x. Use Procedure G16 to graph the two functions and their sum. The graph is in Figure 5.5. The first function graphed is

$y = 2 \sin x$, the second function is $y = 0.5 \cos 6x$, and the third function is the composite function $y = 2 \sin x + 0.5 \cos 6x$.

Exercise 5.3

In Exercises 1-12 use the trigonometric viewing rectangle.

1. Graph each of the following pairs of functions on your graphing calculator. Estimate to one decimal place from the graph (no need to zoom in) the x-intercepts of each graph and the coordinates of any local maximum and/or local minimum points. Give equations of any vertical asymptotes.
   (a) $y_1 = \sin x$ and $y_2 = \csc x$
   (c) $y_1 = \tan x$ and $y_2 = \cot x$
   (e) $y_1 = \cos x$ and $y_2 = \sec x$
   [Recall that $\cot x = \dfrac{1}{\tan x}$, $\sec x = \dfrac{1}{\cos x}$, and $\csc x = \dfrac{1}{\sin x}$.]
2. Graph $y = \sec x$, $y = 2 \sec x$ and $y = 3 \sec x$.
   (a) What is the difference between the graphs of these functions?
   (b) What is the effect of A in the equations $y = A \sec x$?
   (c) By hand, sketch a graph of $y = 5 \sec x$. Check your answer on the calculator.
3. Graph $y = \tan x$, $y = 2 \tan x$ and $y = 3 \tan x$.
   (a) What is the difference between the graphs of these functions?
   (b) What is the effect of A in the equations $y = A \tan x$?
   (c) By hand, sketch a graph of $y = 0.5 \tan x$. Check your answer on the calculator.
4. Graph $y = \cot x$, $y = 2 \cot x$, and $y = 4 \cot x$
   (a) What is the difference between the graphs of these functions?
   (b) What is the effect of A in the equations $y = A \cot x$?
   (c) By hand, sketch a graph of $y = 0.5 \cot x$. Check your answer on the calculator.
5. Graph $y = \tan x$, $y = \tan (2x)$ and $y = \tan(4x)$.
   (a) What is the difference between the graphs of these functions?
   (b) What is the effect of B in the equation $y = \sec Bx$?
   (c) By hand, sketch a graph of $y = \tan (0.5x)$. Check your answer on the calculator.
6. Graph $y = \csc x$, $y = \csc (2x)$, and $y = \csc (0.5x)$,
   (a) What is the difference between the graphs of these functions?
   (b) What is the effect of B in the equations $y = \csc Bx$?
   (c) By hand, sketch a graph of $y = \csc(4x)$. Check your answer on the calculator.

7. Graph $y = 2 \tan x$, $y = 2 \tan (x + \frac{\pi}{4})$, and $y = 2 \tan (x - \frac{\pi}{2})$.
   (a) What is the difference between the graphs of these functions?
   (b) What is the effect of C in the equations $y = 2 \tan (x + C)$?
   (c) By hand, sketch a graph of $y = 4 \tan (x + \frac{\pi}{2})$. Check your answer on the calculator.
8. Graph $y = 3 \cot x$, $y = 3 \cot (x - \frac{\pi}{4})$, and $y = 3 \cot (x + \frac{\pi}{2})$.
   (a) What is the difference between the graphs of these functions?
   (b) What is the effect of C in the equations $y = 2 \cot (x + C)$?
   (c) By hand, sketch a graph of $y = 4 \cot (x + \frac{\pi}{2})$. Check your answer on the calculator.
9. Graph $y = 2 \sec (0.5x)$ and $y = -2 \sec (0.5x)$ on the same graph.
   (a) What is the effect of the negative sign?
   (b) By hand, sketch the graph of $y = -3 \csc x$. Check your answer on the calculator.
10. Graph $y = 2 \tan x$ and $y = -2 \tan x$ on the same graph.
   (a) What is the effect of the negative sign?
   (b) By hand, sketch the graph of $y = -2 \cot x$. Check your answer on the calculator.

♦ 11. *(Washington, Exercises 10-4, #23)* A mechanism with two springs moves according to the equation $b = 4.25 (\sin \frac{\pi}{4}) \csc A$ where A is the angle between the springs and b is the length of one of the springs in centimeters. Graph b as a function of angle A ($0 \le A \le \pi$).
   (a) Determine b when A = 1.00 radian and when A = 2.20 radians.
   (b) What is the minimum value for b and for what A does it occur?
   (c) What happens to b as A gets closer to π?

♦ 12. *(Washington, Exercises 10-4, #22)* At a distance s from the base of a building 215 m high, the angle of elevation θ to the top of the building can be found from the equation $s = 215 \cot \theta$. Graph this function.
   (a) Determine the angle θ when s = 100.0 m and when s = 250 m.
   (b) Does it make any sense to try to determine y when the angle $\theta = 2.0$ radians? Explain your answer.
   (c) What happens to the angle θ as the distance s becomes large?

In problems 13-22, graph y_1, graph y_2, then graph the composite graph.$y_1 + y_2$. Graph one complete cycle of the composite graph for values of x greater than or equal to zero. Determine the local maximum and local minimum points of the composite graph and the x-intercepts of the composite graph to two significant digits.

13. $y = 2, y = \cos x$
14. $y = 0.05x^3, \; y = \sin x$
15. $y = 0.2x^2, y = \cos 2x$
16. $y = x + 2, y = 2 \sin 2x$
17. $y = \sin 3x, y = \cos 2x$
18. $y = \sin 3x, y = \cos 4x$
19. $y = 2 \sin 3x, y = 4 \sin 2x$
20. $y = 3 \cos 0.5x, y = 2 \cos x$
21. $y = 2 \sin (2x + \frac{\pi}{4}), y = 2 \cos 3x$
22. $y = 3 \sin (x - \frac{\pi}{6}), y = 2 \cos (x - \frac{\pi}{3})$
23. A higher frequency cosine wave given by $y = 0.4 \cos 25x$ is superimposed on a lower frequency cosine wave $y = 2 \cos 3x$. Graph one complete cycle of the composite graph.
    (a) Determine the point that gives the largest value of y for x between 1 and 3.
    (b) Determine all x-intercepts between $x = 0$ and $x = 2$.
24. A sine wave given by $y = 2 \sin 100\pi x$ is superimposed on a lower frequency cosine wave given by $y = 10 \cos 10\pi x$. Graph one complete cycle of the composite graph.
    (a) Determine the point that gives the smallest value for y for x between 0 and 2.
    (b) Determine all x-intercepts between $x = 0$ and $x = .25$.

◆ 25. *(Washington, Exercises 10-6, #35)* The vertical displacement of a buoy floating in the water is given by $y = 2.50 \cos 0.24t + 1.2 \sin 0.36t$, where y is measured in feet and t is measured in seconds. Graph one complete cycle of this graph.
    (a) What is the t-value of the first point (for $t > 0$) that the curve reaches its maximum value? What is the maximum value?
    (b) Determine the maximum distance from the lowest point on the wave to the highest point on the wave.

◆ 26. *(Washington, Exercises 10-6, #38)* The available solar energy depends on the amount of sunlight, and the available time in a day for sunlight depends on the time of the year. An approximate correction factor, in minutes, to standard time is $C = 10 \sin \frac{1}{29}(n - 80) - 7.5 \cos \frac{1}{58}(n - 80)$ where n is the number of the day of the year. Graph one complete cycle of this graph.
    (a) Give all the values of n for which the correction factor is zero in the first cycle for $n > 0$.
    (b) For what value of n would the correction factor be at its maximum point?

When certain signals are sent to an oscilloscope special waves may be seen on the oscilloscope screen. These waves may be represented by what is called a Fourier series. Exercises 27 and 28 give the first three terms of two Fourier series. In each case graph the first term of the series, then the sum of first two terms, and then the sum of the first three terms. Guess the next term in the series and graph the sum of the first four terms of the series. On paper sketch the type of wave that the graph approaches as more terms are considered and give the length of one cycle.

27. $y = \sin(\pi x) + \frac{1}{2}\sin(2\pi x) + \frac{1}{3}\sin(3\pi x)$

28. $y = \sin(\pi x) + \frac{1}{3}\sin(3\pi x) + \frac{1}{5}\sin(5\pi x)$

5.4 Graphing of Parametric Equations *(Washington, Section 10-6)*

Parametric equations are equations in which two or more variables are expressed in terms of a third variable called the parameter. For the two dimensional case we consider the variables x and y expressed in terms of a third variable t ($x = f(t)$ and $y = g(t)$). The variable t is called the parameter. The letter t is frequently used for the parameter since in many cases it is appropriate to consider time as the third variable. The variable t is the independent variable and x and y are both dependent variables. As t varies, the values for x and y change accordingly. On a graph the points (x, y) are plotted and connected in order of increasing values of t.

Parametric equations are easily graphed on the graphing calculator after first changing to the correct mode.

Procedure G17. To change graphing modes for regular functions, parametric, or polar equations.

1. Set the graphing mode:
   Press the key MODE.
   Make active:

On the TI-82:	On the TI-85:
(a) For graphing in rectangular coordinates:	
In fourth line: Func	In fourth line: RectC In fifth line: Func
(b) For graphing parametric equations:	
In fourth line: Par	In fourth line: RectC In fifth line: Param

(c) For graphing in polar coordinates:

On the TI-82:	On the TI-85:
In fourth line: Pol	In fourth line: PolarC In fifth line: Pol

Then, press QUIT or CLEAR.

2. Set the graphing format:

On the TI-82:	On the TI-85:
Press the key WINDOW	Press the key GRAPH
Highlight FORMAT in the top row.	From the menu, select FORMT (after pressing MORE)
Then in second line:	Then in first line:

Make active:
(a) For graphing in rectangular coordinates or for parametric equations: RectGC
(b) For graphing in polar coordinates: PolarGC
Then, press QUIT or CLEAR.

(To make an item active, move the cursor to that item and press ENTER.)

After changing to the correct graphing mode, we next enter the parametric equations in the function table in much the same way as we entered regular functions.

Procedure G18. To graph parametric equations.

1. Make sure the calculator is in correct mode for parametric equations. (Procedure G17)
2. To enter the equations:

On the TI-82:	On the TI-85:
Press the key Y=	Press the key GRAPH.
On the screen will appear:	From the menu, select E(t) =
X_{1t}=	On the screen will appear:
Y_{1t}=	xt1=
X_{2t}=	yt1=
Y_{2t}=	(up to 99 pairs of equations may be entered.)
(etc.)	
(up to 6 pairs of equations may be entered)	

Functions are not turned on (the equal signs are not highlighted) until functions have been entered for both x and y. Functions may be turned on and off as with functions in rectangular coordinates (see Note 3, Procedure G1). Only functions turned on will be graphed.

3. Enter the first equation for x in the first row and the corresponding equation of y in the second row. If there are other sets of functions you wish to graph, enter them for the other x's and y's.
To obtain the variable t

On the TI-82:	On the TI-85:
Press the key X,T,θ	Select t from the menu.

4. Press or select GRAPH to see the graph or QUIT to exit.

Some adjustment of the viewing rectangle may be necessary. The setting of x and y values are handled as previously mentioned (Procedures G5 and G6). However, we also need to set values for t.

Procedure G19. To change values for the parameter T.

With the calculator in the mode for parametric equations:

1. Follow Procedure G5 to see the values for the viewing rectangle.
2. Enter new values for Tmin, Tmax, and Tstep and press ENTER or move to the next item by using the cursor keys.
   The standard viewing rectangle values are:

   Tmin = 0, Tmax = 6.28...(2π), and Tstep = 0.13...($\pi/24$)
   (or Tmax = 360 and Tstep = 7.5 if in degree mode).

   Tstep determines how often values for x and y are calculated. The size of Tstep will affect the appearance of the graph. Tmax should be sufficiently large to give a complete graph.
3. Press QUIT to exit.

The step size for T will affect the appearance of the graph. If the step size is too large, a series of line segments may result rather than a smooth appearing curve. A step size of approximately 0.1 is usually sufficient. However, be aware that if we zoom in on a portion of the graph, this step size may be too large.

Example 5.5: Graph the parametric equations $x = 4 \sin 2t$ and $y = 5 \cos (t + \frac{\pi}{4})$ for $0 \le t \le 2\pi$. If the graph has end points, give the coordinates of the end points and estimate the x- and y-intercepts to one decimal place..

Solution:

1. Use procedure G17 to change the mode to graph parametric equations and make sure the calculator is set up for using radians.
2. It is generally a good idea to graph parametric equations using the square viewing rectangle. First obtain the standard viewing rectangle, then the square viewing rectangle (see Procedure G6). This will set Tmin = 0 and Tmax = 2π.
3. The function is defined by :

   $$x = 4 \sin (2t)$$

   $$y = 5 \cos (t + \frac{\pi}{4})$$

   Enter these and graph according to Procedure G18.

4. The graph is drawn in order of increasing values of the variable t starting at $t = 0$ and going to $t = 2\pi$ (if this is the maximum of T). The graph is in Figure 5.5 and appears to be part of a parabola opening to the left. Using the trace function we determine the end points to be at $(-4, -5)$ and $(-4, 5)$. The x-intercept is 4 and the y-intercepts are -3.5 and 3.5.
   Notice that when using the trace function the graph is traced out in the order that t increases or decreases.

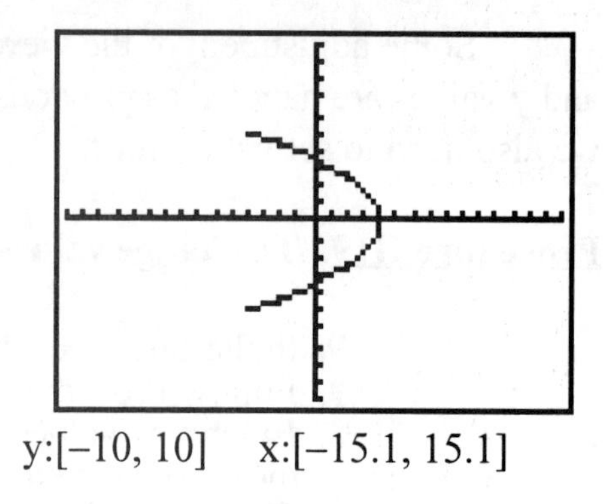

y:[−10, 10] x:[−15.1, 15.1]

Figure 5.6

To obtain a complete graph or to see particular features of a graph it may be necessary to alter the viewing rectangle. If value of t are not given in a problem, select values that give a complete graph. If the graph is cyclic (that is, the same graph is retraced as t increases), select a maximum value of t to give one complete cycle of the graph.

Exercise 5.4

Graph the following sets of parametric equations for the specified values of t. For best results, the calculator should be in radian mode and you should use a square viewing rectangle. (a) Draw a sketch of the complete graph, (b) give the x- and y-intercepts of the graph, and if the graph has end points, give the coordinates of the end points .
Note: The viewing rectangle may have to be changed to see a complete graph.

1. $x = 3t - 5$, $y = 6 - t$, $-3 \leq t \leq 3$
2. $x = 26 + 6$, $y = 3t + x$, $-4 \leq t \leq 4$
3. $x = 3t$, $y = t^2$, $-5 \leq t \leq 5$
   What happens to the graph as A is changed in the equations $x = At$, $y = t^2$?
4. $x = t^2$, $y = t$, $-5 \leq t \leq 5$
   What happens to the graph as B is changed in the equation $x = t^2$, $y = Bt$?
5. $x = 4t$, $y = 4t^{-1}$, $-8 \leq t \leq 8$
6. $x = t$, $y = \sqrt{25 - t^2}$, $-5 \leq t \leq 5$

In problems 7-12, use $0 \le t \le 2\pi$. The graphs obtained are figures that may appear on an oscilloscope. They are called Lissajous figures and are best graphed using a square viewing rectangle. Draw a sketch of the graph obtained and give the x- and y-intercepts.

7. $x = 3 \sin t$, $y = 5 \cos t$
8. $x = 5 \sin t$, $y = 5 \sin t$
9. (a) Graph $x = 3 \sin t$, $y = 5 \sin t$.
   (b) Graph $x = 3 \sin 2t$, $y = 5 \sin t$.
   (c) Graph $x = 3 \sin 4t$, $y = 5 \sin t$.
   What is the effect of changing A in the equations $x = 3 \sin At$, $y = 5 \sin t$? Is there any difference if A is an even or an odd integer?
10. (a) Graph $x = 3 \sin t$, $y = 5 \cos t$.
   (b) Graph $x = 3 \sin 2t$, $y = 5 \cos t$.
   (c) Graph $x = 3 \sin 4t$, $y = 5 \cos t$.
   What is the effect of changing A in the equations $x = 3 \sin At$, $y = 5 \cos t$? Is there any difference if A is an even or an odd integer?
11. (a) Graph $x = 5 \sin t$, $y = 5 \cos t$.
   (b) Graph $x = 5 \sin (t + \frac{\pi}{4})$, $y = 5 \cos t$.
   (c) Graph $x = 5 \sin (t + \frac{\pi}{2})$, $y = 5 \cos t$.
   What is the effect of changing C in the equations $x = 5 \sin (t + C)$, $y = 5 \cos t$? What happens if $C = \pi$?
12. (a) Graph $x = 5 \sin (2t + \frac{\pi}{2})$, $y = 5 \cos 2t$.
   (b) Graph $x = 5 \sin (2t + \frac{\pi}{2})$, $y = 5 \cos 3t$.
   (c) Graph $x = 5 \sin (2t + \frac{\pi}{2})$, $y = 5 \cos 4t$.
   What is the effect of changing B in the equations $x = 5 \sin (2t + \frac{\pi}{2})$, $y = 5 \cos Bt$?
   Does it make a difference if B is even or odd?

Graph the following parametric equations. Use $-4\pi \le t \le 4\pi$. Draw a sketch of the graph obtained.

13. $x = t - \sin t$, $y = 1 - \cos t$ (Called a cycloid)
14. $x = \cot t$, $y = (\sin t)^2$ (Called the Witch of Agnesi)

Chapter 6

Vectors and Complex Numbers

6.1 Working with Vectors.
(Washington, Section 9-2 and 9-3)

Many quantities, such as length or area, may be described by giving a single number corresponding to the magnitude of that quantity. However, other quantities such as force, velocity, and acceleration are best described giving both the magnitude and the direction. In mathematics, **vectors** are quantities that can be described by giving both magnitude and direction and are represented as directed line segments. To show a vector we to draw a line of a certain length, and, also, to place an arrow on the end of the line to show its direction. To describe a vector, we need to give a number indicating the magnitude of the vector and an angle indicating the direction of the vector.

For example, suppose we wish to show a vector **F** describing a force of 20.0 pounds acting in a direction of 40 degrees above the horizontal. We would first select a scale (perhaps 1 cm = 5 lb), and draw a line segment of the appropriate length and at an angle of 40 degrees as in Figure 6.1. An arrow is drawn at the end of the line segment to show the direction of the vector.

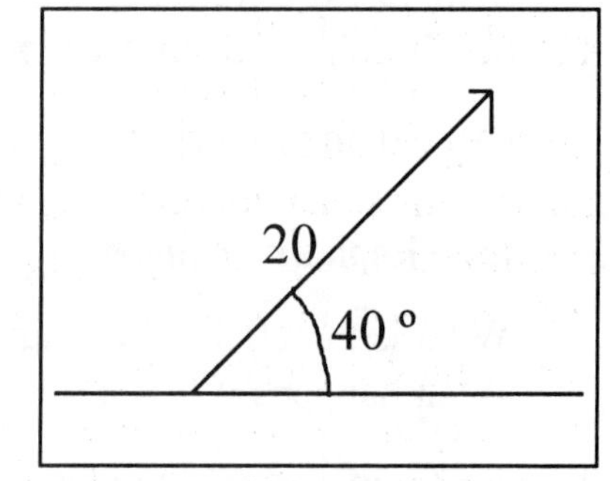

Figure 6.1

Vectors are often named using letters which are printed in bold type to denote that it is a vector (for example, **A**). When writing vectors by hand the name of the vector should be written with an arrow above it (for example, $\vec{A}$). Writing the name of the vector without the arrow or printing the name in regular type represents the magnitude of the vector. The Greek letter θ is often used to represent the angle to the vector. Often the name of the vector is written as a subscript to θ to denote the angle to a particular vector (for example θ_A). For the vector **F**, $F = 20.0$ and $\theta_F = 40°$.

It is frequently helpful to break down the vector into horizontal and vertical components. These components are vectors in the horizontal and vertical direction whose sum is the given vector.

It is particularly advantageous to consider vectors relative to the x- and y-coordinate system. Vectors are placed to start at the origin and the direction of the vector is the angle to the vector measured from the positive x-axis. Such a vector is said to be in **standard position**. The horizontal component is then the x-component and the vertical component is the y-component. The x-component is a vector that starts at the origin and is directed along the x-axis and the y-component starts at the origin and is directed along the y-axis. For a vector **V**, the magnitude of the x-component is designated as V_x and the magnitude of the y-component is designated as V_y. If we let θ represent the angle to the vector from the positive x-axis, then these components are given by

$$V_x = V \cos \theta \qquad \text{[Equation 1]}$$
$$V_y = V \sin \theta$$

For vector **F** above, the magnitude of the horizontal (or x-) component is given by 20.0 cos 40° = 15.3 and of the vertical (or y-) component is given by 20.0 sin 40° = 12.9.

In order to have a convenient method of working with vector components on a graphing calculator, we next give a brief introduction to the location of points on a graph using polar coordinates. (Polar coordinates and graphing in polar coordinates will be covered in more detail in Chapter 11.) In rectangular coordinates, a point is identified by giving the x- and y-coordinates of the point. In polar coordinates, a point is identified by giving the distance, r, of the point from the origin and the angle, θ, from the positive x-axis to the line going from the origin to the point. Polar coordinates are expressed in the form (r, θ). Any point can be identified by giving either the rectangular coordinates of that point or the polar coordinates of the point. In order to convert from one set of coordinates to the other, we use the relationships

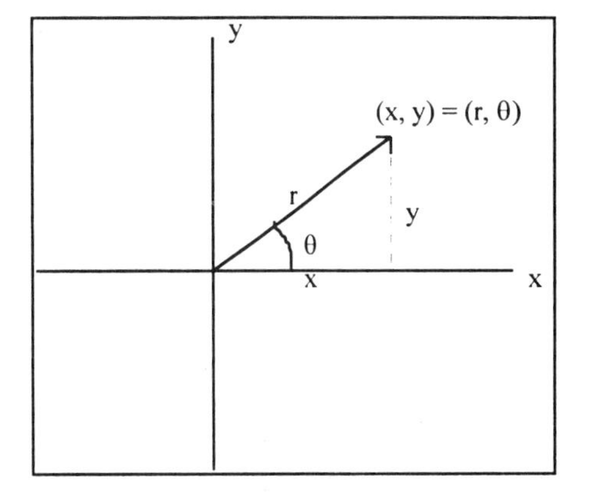

Figure 6.2

$$x = r \cos \theta \quad \text{and} \quad r = \sqrt{x^2 + y^2} \qquad \text{[Equation 2]}$$
$$y = r \sin \theta \qquad \theta = \tan^{-1} \frac{y}{x}$$

Let **V** be a vector that begins at the origin and terminates at the point (x, y) and let (r, θ) be the polar coordinates of this point. Then, the magnitude of the vector equals the r-value and the direction of the vector equals the angle θ. Finding the components of the vector from its magnitude and direction is equivalent to converting polar coordinates to rectangular coordinates. Likewise, finding the magnitude and direction of a vector from the components is equivalent to converting rectangular coordinates to polar coordinates.

Vector	Polar coordinates	Rectangular coordinates
magnitude	r	$\sqrt{x^2 + y^2}$
direction	θ	$\tan^{-1} \frac{y}{x}$
x-component	r cos θ	x
y-component	r sin θ	y

For our purposes, a vector in terms of its x- and y-components will considered in **rectangular form** and be expressed as [x, y]. If the vector is in terms of its magnitude and direction, it will be considered to be in **polar form** and expressed as [r ∠θ].

Many graphing calculators contain a provision for converting from polar to rectangular coordinates and from rectangular to polar coordinates. This feature may be used to find the components of vectors given the magnitude and direction and to find the magnitude and direction given the components.

Procedure C15. Conversion of Coordinates

Make sure calculator is in correct mode -- radians or degrees.

A. To change from polar to rectangular:

On the TI-82:

1. Press the key ANGLE
2. To find x:
   Select 7:P▷Rx(
   To find y:
   Select 8:P▷Ry(
3. Enter the values of r and θ separated by commas and press ENTER.

Example: P▷Rx(2.0, 45)
gives 1.4 [in degree mode]

On the TI-85:

1. Enter the magnitude and angle separated by a ∠ in [ ].
2. Press the key VECTR
3. Select OPS
4. Select ▷Rec (after pressing MORE)
5. Press ENTER

Example: [2.0∠45]▷Rec
gives [1.4, 1.4]

B. To change from rectangular to polar:

On the TI-82:

1. Press the key ANGLE
2. To find r:
   Select 5:R▷Pr(
   To find θ:
   Select 6:R▷Pθ(
3. Enter the values of x and y separated by commas and press ENTER.

Example: R▷Pθ(1.0, 1.0)
gives 45 [in degree mode]

Note: On the TI-82, it may be just as easy to use the conversion formulas (Equation 2) directly -- especially when converting from polar to rectangular.

On the TI-85:

1. Enter the values of x and y inside of [ ] separated by a comma.
2. Press the key VECTR
3. Select OPS
4. Select ▷Pol
5. Press ENTER

Example: [1.0, 1.0] ▷Pol
gives [1.4 ∠45]

Note: On the TI-85, vectors are entered or displayed as a pair of numbers inside of square brackets, [ ]. When the numbers are separated by a comma they are in rectangular form and when separated by ∠ are in polar form.

After conversion from one form to the other, a visual check should be made to make sure the two forms represent the same vector. When converting from rectangular to polar, be careful to obtain the correct angle. When the angle is displayed as a negative angle, to obtain the corresponding positive angle, add the negative angle to 360°.

Example 6.1: (a) Find the x- and y-components of the vector with a magnitude of 3.25 at an angle of 150°.
(b) Find the magnitude and angle for a vector that has an x-component of 2.14 and a y-component of −3.56.

Solution:

1. Finding the x- and y-components in part (a) is equivalent to converting from polar coordinates to rectangular coordinates. The x-component may be found using 3.25 cos 150° and the y-component may be found using 3.25 sin 150° or use Procedure C15 to perform this conversion. The x-component of this vector is −2.81 and the y-component is 1.63. The answer is in rectangular form which we write as [−2,81, 1.63].
2. Finding the magnitude and direction in part (b) is equivalent to converting from rectangular coordinates to polar coordinates. Use Procedure C15 and perform this conversion. The magnitude is 4.15 and the angle is given as −59.0 which is the same as 301.0°. The answer is in polar form which we write as [4.15 ∠301°]

The sum of two or more vectors is referred to as the **resultant** of the vectors. In order to add vectors in polar form, first find the x- and y-components of each vector. Then add all the x-components and all the y-components to obtain the x-component and y-component of the resultant vector. The magnitude and direction of the resultant vector are then obtained from the components of the resultant vector. On some calculators it may be possible to simply enter the vectors in polar form and add the vectors without finding the components.

Procedure C16. Finding the sum of vectors in polar form.

On the TI-82:

1. Find the x- and y-components of each vector. (see Procedure C15)
2. Find the sum of the x-components and the sum of the y-components.
3. Find the magnitude and direction of the resultant vector.
   (see Procedure C15)

or

Find the sum

$A\cos\theta_A + B\cos\theta_B + \ldots$

and store this for x and find

$A\sin\theta_A + B\sin\theta_B + \ldots$

and store this for y,
then convert [x, y] to polar form.

On the TI-85:

Vectors may be entered directly in polar form and the added together.

Example:

[12 ∠120] + [15 ∠75]
gives [25 ∠95]

Note: On the TI-85, vectors may be stored under variable names by entering the vector, pressing the key STO▷, entering the variable name and pressing ENTER.
Vectors may also be entered or modified by first pressing the key VECTR, then selecting EDIT.
The sum of vectors may be found by finding the sum of the variables.

The form of the output of a vector on the screen may be changed by pressing the key MODE and in the seventh line highlighting

RectV for rectangular form
CylV for polar form

Example 6.2: Find the sum of the vectors A and B where A has a magnitude of 22.5 in the direction $\theta_A = 75.8°$ and B has a magnitude of 57.2 in the direction $\theta_B = 153.2°$.

Solution: Use Procedure C16 to find the sum of the two vectors.
We find that $[22.5 \angle 75.8] + [57.2 \angle 153.2] = [65.9 \angle 133.7]$
Thus, the sum has a magnitude of 65.9 and is in the direction of 133.7°.

Exercise 6.1

In Exercise 1-8, each vector is given in polar form, find the x- and y-components of the vector. The given vectors are in standard position.

1. magnitude = 121, angle = 87.2°
2. magnitude = 35.7, angle = 125.6°
3. magnitude = 0.175, angle = 227.6°
4. magnitude = 1273, angle = 312.5°
5. $[12243 \angle 172.8°]$
6. $[65.3 \angle 270.0°]$
7. $[1.25 \angle 89.8°]$
8. $[789 \angle 157.5°]$

In Exercises 9-16, each vector is given in rectangular form, find the magnitude and direction of the vector. Each vector is in standard position.

9. x-component is 16.5, y-component is 25.2
10. x-component is 4275, y-component is −3672
11. x-component is −1.475, y-component is −0.1758
12. x-component is −17.3, y-component is 37.4
13. [1273, 127.3]
14. [−6.72, 0.00]
15. [−152, 43.5]
16. [16530, −87634]

In Exercises 17-26, find the sum, $\vec{A} + \vec{B}$ or $\vec{A} + \vec{B} + \vec{C}$, of the given vectors.

17. $A = 24.7$, $\theta_A = 37.8°$ and $B = 12.5$, $\theta_B = 115.9°$
18. $A = 125.2$, $\theta_A = 85.64°$ and $B = 312.8$, $\theta_B = 212.35°$
19. $\vec{A} = [52.34 \angle 134.34°]$ and $\vec{B} = [37.25 \angle 226.46°]$
20. $\vec{A} = [2.17 \angle 12.2°]$ and $\vec{B} = [3.48 \angle 294.6°]$
21. $A = 17.8$, $\theta_A = 35.4°$, $B = 25.2$, $\theta_B = 157.6°$, and $C = 16.1$, $\theta_C = 270.0°$
22. $A = 1.752$, $\theta_A = 112.2°$, $B = 2.175$, $\theta_B = 180.0°$, and $C = 5.143$, $\theta_C = 345.6°$
23. $\vec{A} = [243.1 \angle 16.35°]$, $\vec{B} = [1724.4 \angle 85.63°]$, and $\vec{C} = [1934.6 \angle 220.64°]$
24. $\vec{A} = [12370 \angle 184.3°]$, $\vec{B} = [11540 \angle 269.5°]$, and $\vec{C} = [24310 \angle 169.3°]$

25. The force acting on an object is 24.3 pounds at an angle of 42.3° above the horizontal. Find the horizontal and vertical components.
26. The force on a cable is 5674 pounds at an angle of 152.5°. Find the horizontal and vertical components.
◆ 27. *(Washington, Exercises 9-2, #21)* A jet is 145.5 km at a position 37.35° north of east of a city. What are the components of the jet's displacement from the city?
◆ 28. *(Washington, Exercises 9-2, #18)* Water is flowing downhill at 17.75 ft/s through a pipe that is at an angle of 65.68° with the horizontal. What are the horizontal and vertical components of the velocity?
29. Two forces act on a point. One force is 125 lb at and angle of 72.3° with respect to the horizontal and the other is 112 lb at and angle of 312.4° to the horizontal. Find the resultant force.
30. A plane is heading directly toward the northeast at 625 km/h and the wind is blowing directly toward the southeast at 65 km/h. What is the actual velocity and direction of the plane (the resultant of these two vectors)?

6.2 Complex Numbers
(Washington, Sections 12-2, 12-4, and 12-6)

If we define $j = \sqrt{-1}$, then a **complex number** is any number that can be written in the form a + bj where a and b are real numbers. Since a complex number is represented by two real values, each complex number corresponds to a point on the rectangular coordinate system where values on the x-axis correspond to the real part of the complex number and values on the y-axis correspond to the imaginary part of the complex number. For the complex number 3 + 2j, the real part, 3, corresponds to and x-coordinate 3 and the real number, 2, in the imaginary part, corresponds to the y-coordinate 2. Then the complex number 3 + 2j can be represented by a vector from the origin to the point (3, 2). A coordinate system set up in this way for complex numbers is called the **complex plane** and each complex number corresponds to a vector. In the complex plane, the x-axis is the called the **real axis** and the y-axis is called the **imaginary axis**.

Considering the complex number x + yj as a vector from the origin to the point (x, y) with a magnitude r and a direction θ, we have the familiar relationships or x, y, r and θ given in Equation 2. Using $x = r \cos \theta$ and $y = r \sin \theta$, the complex number x + yj may also be written in the form $r \cos \theta + (r \sin \theta)\, j$ or the form $r(\cos \theta + j \sin \theta)$. The form x + yj is called the **rectangular form** of the complex number and the form $r(\cos \theta + j \sin \theta)$ is called the **polar form**. Another notation for $r(\cos \theta + j \sin \theta)$ is $r \angle\theta$.

A vector with x- and y-components corresponds to the rectangular form of a complex number x + yj. A vector identified by its magnitude and direction corresponds to the polar form of a complex number $r\angle\theta$. Finding the sum of two complex numbers is similar to finding the sum of two vectors (see Procedure C16).

All of the familiar operations of addition, subtraction, multiplication, and division may be performed on complex numbers. For complex numbers in rectangular form, these operations are similar to algebraic operations as shown in the following examples. Recall that since $j = \sqrt{-1}$, then $j^2 = -1$.

$(2 + 3j) + (3 - 4j) = 5 - j$
$(2 + 3j) - (3 - 4j) = -1 + 7j$
$(2 + 3j)(3 - 4j) = 6 - 8j + 9j - 12j^2 = 18 + j$

For division, multiply the numerator and denominator by the conjugate of the denominator. The **conjugate** of the number a + bj is a – bj.

$$\frac{2+3j}{3-4j} = \frac{(2+3j)(3+4j)}{(3-4j)(3+4j)} = \frac{6+8j+9j+12j^2}{9-16j^2} = \frac{-6+17j}{25} = -\frac{6}{25} + \frac{17}{25}j$$

The **absolute value of a complex number** is the magnitude of the corresponding vector:

$$|2+3j| = \sqrt{2^2+3^2} = \sqrt{13}$$

If the complex numbers are given in polar form, then the operations may be performed by first converting the complex number to rectangular form, then proceeding to do the operations in rectangular form. The result can then be converted back to polar form. Some calculators perform operations on complex numbers directly regardless of the form of the complex numbers.

	vectors	complex numbers
rectangular form	[x, y]	x + yj or (x, y)
polar form	[r ∠θ]	r ∠θ or (r ∠θ)

Procedure C17. Conversion of complex numbers.

Make sure calculator is in correct mode -- radians or degrees.

A. To change from polar to rectangular form:

On the TI-82:
Think of the complex number r ∠θ as r and θ in polar form and use Procedure C15.

On the TI-85:
1. Enter the magnitude and angle separated by a ∠ in ( ).
2. Press the key CPLX
3. Select ▷Rec (after pressing MORE) and press ENTER

Example: (2.0 ∠45)▷Rec
gives (1.4, 1.4)

B. To change from rectangular to polar form:

On the TI-82:
Think of the complex number x + yj as x and y in rectangular form and use Procedure C15.

On the TI-85:

1. Enter the values of x and y inside of ( ) separated by a comma.
2. Press the key CPLX
3. Select ▷Pol
4. Press ENTER

Example: (1.0, 1.0) ▷Pol gives (1.4 ∠45)

Note: On the TI-85, a complex number is expressed as a pair of numbers inside of parentheses ( ). When the numbers are separated by a comma they are in rectangular form and when separated by ∠ they are in polar form.

Example 6.3: (a) Convert 12.4(cos 25.8° + j sin 25.8°) to rectangular form.
(b) Convert the complex number 3.5 – 2.7j into polar form.

Solution:

1. In part (a), the complex number 12.4(cos 25.8° + j sin 25.8°) = 12.4 ∠25.8°. Use Procedure C17 to convert this to rectangular form and obtain 11.2 + 5.39j.
2. For part (b) use Procedure C17 to convert the complex number 3.5 – 2.7j to polar form. The result is 4.4 ∠322°.

Procedure C18. Operations on complex numbers.

On the TI-82:

1. If not already in rectangular form, change the complex number into rectangular form as when changing polar coordinates to rectangular coordinates (see Procedure C15).
2. Perform the operation on the rectangular forms of the complex numbers.

Operations are performed by entering the complex numbers in either rectangular or polar form using the operation symbols +, –, ×, or ÷ between the numbers.

Note: Powers and roots of complex number may be found by raising the complex number to the desired power using ^.

3. If the answer is to be in polar, convert the result back to polar form as when changing rectangular coordinates to polar coordinates (see Procedure C15).

The form of output of the complex number on the screen may be changed by pressing the key MODE and in the fourth line highlighting
RectC for rectangular mode
PolarC for polar mode

Example 6.4: Find the following and give answers in polar form:
(a) $2.24 \angle 135.4° + 3.18 \angle 275.7°$
(b) $(2.24 \angle 135.4°)(3.18 \angle 275.7°)$

Solution:
1. In part (a), use Procedure C18 to find the sum of the two complex numbers. The result is $2.04 \angle 231.2°$.
(The rectangular form of $2.24 \angle 135.4°$ is $-1.59 + 1.57j$ and the rectangular form of $3.18 \angle 275.7°$ is $0.316 - 3.16j$.)
2. In part (b), use Procedure C18 to find the products of the two complex numbers. The result is $7.12 \angle 51.1°$.

Exercise 6.2

In Exercises 1-6, convert the complex number in rectangular form into a complex number in polar form.

1. $3.5 + 2.8j$
2. $-12.4 - 18.6j$
3. $-123.7 + 156.4j$
4. $-67.8j$
5. $-0.125 - 0.0615j$
6. $1275 - 1647j$

In Exercise 7-12, convert the complex number in polar form into a complex number in rectangular form.

7. $837(\cos 121.5° + j \sin 121.5°)$
8. $-12.9(\cos 195.7° + j \sin 195.7°)$
9. $1.654 \angle 227.8°$
10. $3.59 \angle 99.9°$
11. $-87.9 \angle 335.6°$
12. $175.6 \angle 270.0°$

In Exercises 13-24, perform the indicated operation on the complex numbers. Answer should be in same form as the complex numbers in the problem.

13. $(12.5 + 17.8j) + (13.1 - 25.4j)$
14. $(6.134 - 5.234j) - (7.154 + 8.451j)$
15. $(-112.5 + 54.6j)(104.5 - 38.5j)$
16. $\dfrac{2.384 + 1.783j}{1.879 - 3.183j}$
17. $\dfrac{63.8 - 12.4j}{43.8 + 31.4j}$
18. $(0.165 - 1.285j)(-0.341 - 0.897j)$
19. $(25.4 \angle 147°) - (16.3 \angle 226.5°)$
20. $(1.45 \angle 235.9°) + (4.17 \angle 100.4°)$
21. $(-177.2 \angle 178.6°) + (239.4 \angle 85.7°)$
22. $(16.88 \angle 181.32°)(55.32 \angle 289.45°)$
23. $110.2 (\cos 12.4° + j \sin 12.4°) \times 89.3 (\cos 110.4° - j \sin 110.4°)$
24. $72.3 (\cos 127.8° - j \sin 127.8°) + (-55.8)(\cos 226.8° + j \sin 226.8°)$
25. (a) Find $(2 \angle 120)^3$ and $(2 \angle 240)^3$.
    (b) Are $2 \angle 120$ and $2 \angle 240$ cube roots of 8? Explain. Write in rectangular form.
26. (a) Find $(1 \angle 90)^4$ and $(1 \angle 270)^4$.
    (b) Are $1 \angle 90$ and $1 \angle 270$ fourth roots of 1? Explain. Write in rectangular form.

Exercises 27-30 refer to alternating current circuits. In an alternating current circuit the voltage E is given by $E = IZ$ where I is the current in amperes and Z is the impedance in ohms.

◆ 27. *(Washington, Exercises 12-2, #61)* Find the complex number representation of E if $I = 0.867 - 0.524j$ and $Z = 240 + 175j$.

◆ 28. *(Washington, Exercises 12-2, #62)* Find the complex number representation of Z if $E = 6.24 - 4.17j$ and $I = 0.427 - 0.573j$.

29. Find the complex number representation of E and write in rectangular form if $I = 6.25 \angle 58.0°$ and $Z = 85.5 \angle 35.5°$

30. Find the complex number representation of Z and write in rectangular form if $E = 105 \angle 30.5°$ and $I = 4.24 \angle 15.0°$

◆ 31. *(Washington, Exercises 12-6, #47)* In an alternating current circuit the power P in watts is given by $P = EI$ where E is the voltage in volts and I is the current in amperes. Find the complex number representation for P if $E = 7.15 \angle 59.5°$ and $I = 6.35 \angle -14.8°$.

◆ 32. *(Washington, Exercises 12-6, #48)* The displacement d of a weight suspended on a system of two springs is $d = 6.24 \angle 23.4° + 2.88 \angle 75.5°$ inches. Perform the addition and express the answer in polar form.

Chapter 7

Exponent and Logarithm Functions

7.1 Graphing Exponent and Logarithm Functions.
(Washington, Section 13-2)

Exponent and logarithm functions play important role in mathematics. In this section we study these functions by studying their graphs. The graphing of exponent and logarithm functions is basically handled in the same manner as the graphing of algebraic functions. Care does need to be taken to be sure the viewing rectangle is appropriate.

Exponent functions are function that have a variable contained in the exponent. A basic exponent function in the real number system is a function of the form $y = b^x$ where b is a positive number not equal to 1.

Exponent functions and logarithm functions are closely related. From the definition of logarithm it is known that the $\log_b x$ is the power to which we must raise the base number b to obtain the value x. (The value of $\log_2 8$ is 3 since $2^3 = 8$.) The basic logarithm function is of the form $y = \log_b x$. In order for $\log_b x$ to be a real number, x must be positive and b is a positive number not equal to 1.

On the graphing calculator there are two keys for logarithms. One is labeled LOG and the other LN. The LOG key refers to a logarithm of base 10, normally called **common logarithms**, and the LN key refers to logarithms of base e, normally referred to as **natural logarithms**. Thus, $\log x = \log_{10} x$ and $\ln x = \log_e x$. The number **e** is the very important mathematical constant which is approximated by the value 2.71828... (see problem 63, Exercise 1.1). To see the value for e on your calculator, press the key e^x, then press the key 1, then press ENTER.

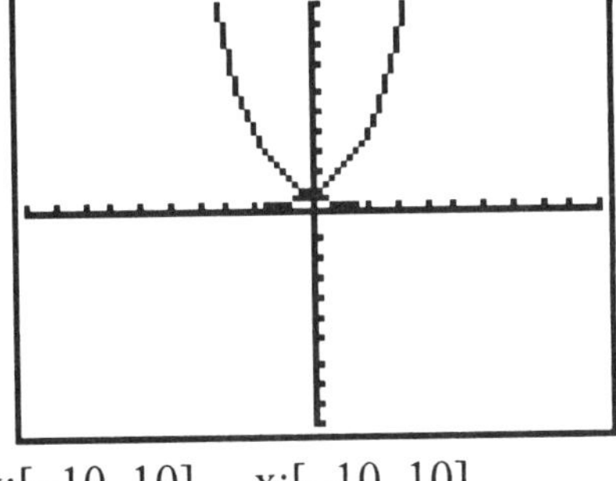

y:[−10, 10]   x:[−10, 10]

Figure 7.1

Example 7.1: Graph the exponent functions $y = 2^x$ and $y = 2^{-x}$ and note how the graph of the second function differs from the graph of the first.

Solution:

1. Using the standard viewing rectangle and Procedure G1 graph the two functions on the same

graph on the calculator. Use the exponent key and enter 2^x for the first function and 2^{-x} for the second function. The resulting graph appears in Figure 7.1.

2. The graph of $Y_1 = 2^x$ is always an increasing function. That is, y becomes larger as x becomes larger. Notice that as x becomes larger than 1, the y values increase more rapidly. The graph of $Y_2 = 2^{-x}$ is always a decreasing function since as x becomes larger, y becomes smaller. In both cases the y-values are always positive and may get close to zero but never become equal to zero.

Recall that an **asymptote** is a line which a graph gets closer and closer to as the distance from the origin becomes greater. In Example 7.1 the x-axis is an asymptote for both graphs.

In order to graph logarithm functions to bases other than e or 10, we use the equation

$$\log_b x = \frac{\log_a x}{\text{lob}_a b}. \qquad \text{[Equation 1]}$$

Either common logarithms or natural logarithms may be used on the right hand side by letting a = 10 or letting a = e. This gives the two equations either of which may be used when working with logarithms in other bases.

If a = 10: $\log_b x = \frac{\log x}{\log b}$ and If a = e: $\log_b x = \frac{\ln x}{\ln b}$

◆ Example 7.2: *(Washington, Section 13-2, Example 6)* Graph the function $y = \log_3 x$ and $y = 3^x$ on the same graph and compare the graphs.

Solution:

1. In this case it is best to have the x- and y-scales the same. Use Procedure G6 to obtain the square viewing rectangle.
2. To enter the function $\log_3 x$ use of the relation $\log_3 x = \frac{\log x}{\log 3}$. Enter log x/log 3 for the first function and 3^x for the second function and graph the functions.

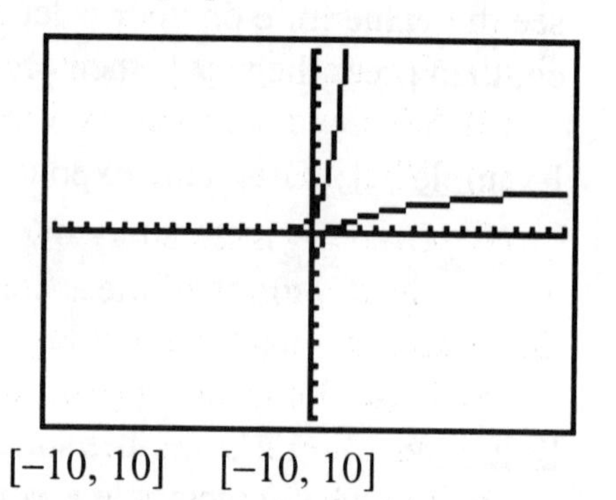

[−10, 10] [−10, 10]

Figure 7.2

The result appears in Figure 7.2.

The graph of $\log_3 x$ is graphed first and is an increasing function for all x. It has have an x-intercept of 1, the domain of this function is the real numbers greater than zero, and the y-axis is an asymptote. The graph of $y = 3^x$ is also an increasing function for all x-values, it crosses the y-axis at 1 and the x-axis is an asymptote.

The two functions in Example 7.2 exhibit an interesting property. Imagine these graphs on a piece of paper on which the line $y = x$ also appears. (The graph of $y = x$ can be added to the graph on your calculator.) If the paper is folded along the line $y = x$ and the x- and y-scales are the same (a square viewing rectangle), the graph of $y = \log_3 x$ will fall on top of the graph of $y = 3^x$. Two functions which have this property are said to be **inverse functions** of each other. This is true of any logarithm function $y = \log_b x$ and its corresponding exponent function $y = b^x$.

Exercises 7.1

1. Graph the functions $y = 2^x$, $y = 3^x$, and $y = 4^x$.
   (a) Describe what happens to the graph as the base number of the exponent function is increased.
   (b) On what intervals are the functions increasing?, decreasing?
   (c) What are the asymptotes of these graphs?
   (d) What are the x- and y-intercepts (if any)?
   (e) What are the domains and ranges of these functions?
   (f) On paper sketch by hand the graph of $y = 5^x$ and determine the equation of the asymptote. Check your answer on the calculator.
2. Graph the functions $y = 2^x - 1$, $y = 2^x$, and $y = 2^x + 2$.
   (a) Describe what happens to the graph as a constant is added to $y = 2^x$.
   (b) On what intervals are the functions increasing?, decreasing?
   (c) What are the asymptotes of these graphs?
   (d) What are the x- and y-intercepts (if any)?
   (e) What are the domains and ranges of these functions?
   (f) On paper sketch by hand the graph of $y = 2^x + 1$ and determine the equation of the asymptote. Check your answer on the calculator.
3. Graph the functions $y = \log_2 x$, $y = \log_3 x$, and $y = \log_4 x$.
   (a) Describe what happens to the graph as the base number of the logarithm function is increased.
   (b) On what intervals are the functions increasing?, decreasing?
   (c) What are the asymptotes of these graphs?

(d) What are the x- and y-intercepts (if any)?
(e) What are the domains and ranges of these functions?
(f) On paper sketch by hand the graph of $y = \log_6 x$ and determine the equation of the asymptote. Check your answer on the calculator.

4. Graph the functions $y = -2 + \log x$, $y = \log x$, and $y = 4 + \log x$.
(a) Describe what happens to the graph as a number is added to the logarithm function.
(b) On what intervals are the functions increasing?, decreasing?
(c) What are the asymptotes of these graphs?
(d) What are the x- and y-intercepts (if any)?
(e) What are the domains and ranges of these functions?
(f) On paper sketch by hand the graph of $y = 2 + \log x$ and determine the equation of the asymptote. Check your answer on the calculator.

5. Graph the functions $y = 4^x$ and $y = 4^{-x}$.
(a) What is the difference between these two graphs?
(b) On what intervals are the functions increasing?, decreasing?
(c) What are the asymptotes of these graphs? Are they the same?
(d) What are the x- and y-intercepts (if any)? Are they the same?
(e) What are the domains and ranges of these functions?
(f) Is one graph the reflection of the other about an axis? If so, which axis?

6. Graph the functions $y = \log x$ and $y = -\log x$.
(a) What is the difference between these two graphs?
(b) On what intervals are the functions increasing?, decreasing?
(c) What are the asymptotes of these graphs? Are they the same?
(d) What are the x- and y-intercepts (if any)? Are they the same?
(e) What are the domains and ranges of these functions?
(f) Is one graph the reflection of the other about an axis? If so, which axis?

7. By hand, sketch graphs of each of the following. Check answers on your calculator.
(a) $y = 8^x$
(b) $y = 3^x + 2$
(c) $f(x) = 2^x + 4$
(d) $g(x) = (\frac{1}{2})^x$ (Note: $(\frac{1}{2})^x = 2^{-x}$)

8. By hand, sketch graphs of each of the following. Check answers on your calculator.
(a) $y = \log_5 x$
(b) $y = 3 + \log x$
(c) $f(x) = -\log_3(x-1)$
(d) $g(x) = -4 + \log_5(x+3)$

For Exercises 9-12, use a square viewing rectangle.

9. Graph the function $y = \log_2 x$ on the calculator. Guess what the graph of the function $y = 2^x$ looks like and sketch it by hand on a piece of paper. Check your answer on the calculator.

10. Graph $y = \ln x$ and $y = e^x$. How are these functions related?
11. Graph $y = \log(x + 2)$ and $y = 10^x - 2$ on the same graph. How are these functions related?
12. Graph $y = 4 + \log_3 x$ and $y = 3^{x-4}$ on the same graph. How are these functions related?

♦ 13. *(Washington, Exercises 13-2, #25)* If an amount of P dollars is invested at an annual interest rate r (expressed as a decimal), the value V of the investment after t years is given by $V = P(1 + \frac{r}{n})^{nt}$, if interest is compounded n times a year. If \$1225 is invested at an annual interest rate of 4.5%, compounded semiannually, express V as a function of t and graph for $0 \le t \le 18$ years. Then, from the graph
   (a) Determine the value after 6.5 years and after 9.0 years.
   (b) Determine the time when the investment will have doubled (when V = \$2450).

14. Repeat Exercise 12 for an annual interest rate of 5.5%.

♦ 15. *(Washington, Exercises 13-2, #27)* Considering air resistance and other conditions, the velocity (in m/s) of a certain falling object is given by $v = 95(1 - e^{-0.1t})$, where t is the time of fall in seconds. Graph this function and from the graph
   (a) Determine the velocity after 8.5 s and after 15 s.
   (b) Estimate what happens as t becomes large. (Hint: let Xmax = 50) Does this graph have an asymptote? What is the limiting velocity (the largest possible value for v)?

♦ 16. *(Washington, Exercises 13-2, #30)* An original amount of 126 mg of radium radioactively decomposes such that N mg remain after t years. The function relating t and N is $t = 2500(\ln 126 - \ln N)$. Graph this function and from the graph
   (a) Determine the time until 100 mg remain and until 63 mg remain.
   (b) Determine how much remains after 600 years after 15000 years.

7.2 Logarithmic Calculations
(Washington, Sections 13-4 and 13-5)

To find logarithms of numbers on a calculator, we make use of the keys LOG and LN to find logarithms to base 10 and base e. Logarithms using other bases may be found by using the Equation 1 in Section 7.1. Example 7.3 shows how the calculator is used to obtain common or natural logarithm values for any positive number.

The logarithm of a number on the calculator may be displayed with several digits. To achieve approximately the same accuracy the number of decimal places in the logarithm of a number should equal the number of significant digits in the number.

Example 7.3: Find log 27.2 and ln 6.254

Solution:
1. To find log 27.2, enter exactly as shown. Press the key LOG, enter 27.2, and press ENTER. The result is log 27.2 = 1.435. (Since 27.2 has 3 significant digits, we round to 3 decimal places.)
2. To find ln 6.254, enter exactly as shown. Press the key Ln, enter 6.254, and press ENTER. The result is ln 6.254 = 1.8332 (Since 6.254 has 4 significant digits, we round to 4 decimal places.)

From the definition of logarithm, log 27.2 = 1.435 means that $10^{1.435} = 27.2$ (approximately). This can be verified by using the exponent key on the calculator to find $10^{1.435}$. This will not give exactly 27.2 because of rounding errors. Likewise, ln 6.254 = 1.8332 means that $e^{1.8332} = 6.254$ (approximately).

Example 7.4: Find $\log_6 7.53$.

Solution: 1. Using common logarithms the equation becomes

$$\log_6 7.53 = \frac{\log 7.53}{\log 6}$$

Enter the right side into the calculator as LOG 7.53 ÷ LOG 6
2. Press ENTER. The result shows that $\log_6 7.53 = 1.127$.

Just as important as finding the logarithm of a number is the procedure for finding a number if we know a logarithm. For example, we may wish to find x given that log x = 2.3145. This procedure is often referred to as finding the antilogarithm. The fact that $x = b^y$ is equivalent to $\log_b = y$ provides a means for finding the antilogarithm. If N is a known number and if log x = N, then from the definition of logarithm, it must be true that $x = 10^N$. Also, if ln x = N, then $x = e^N$. On the calculator the value for 10^N or e^N are given as secondary functions on the LOG and LN keys. For other bases we raise the base to the appropriate power. If $\log_4 x = N$, then $x = 4^N$.

Example 7.5: If ln x = 2.345, find x.

Solution: Since ln x = 2.345 means that $e^{2.345} = x$, evaluate $e^{2.345}$ by using the e^x key. Since the logarithm has 3 decimal places we round the answer to 3 significant digits and the result is x = 10.4.

There are many applications in science and mathematics that require the use of logarithms. One application involves the solving of exponent equations. This makes use of the fact that since the two sides of an equation are equal, the logarithms of the sides are equal. Other applications include equations of the form $d = \log \frac{I}{I_0}$ and the calculation of pH in Chemistry.

Example 7.6: In chemistry, the pH of a solution is given by pH = –log h where h represents the hydrogen ion concentration. Find h if pH = 3.20.

Solution:
1. If pH = 3.20 then 3.20 = –log h, or log h = –3.20. This means that $h = 10^{-3.20}$.
2. Evaluate $10^{-3.20}$. The result is $h = 6.3 \times 10^{-4}$ or 0.00063.

Thus, the hydrogen ion concentration is 0.00063.

Exercises 7.2

In Exercises 1-12, find the required logarithms.

1. $\log 6.35$
2. $\log 0.001378$
3. $\log 16253$
4. $\log 7.5 \times 10^8$
5. $\ln 1.73$
6. $\ln 256350$
7. $\ln 0.06154$
8. $\ln 7.83 \times 10^{15}$
9. $\log_3 17.85$
10. $\log_8 0.0275$
11. $\log_5 1.234 \times 10^6$
12. $\log_{12} 1.024 \times 10^{-7}$

In Exercises 13-22, find the number N, x, or y.

13. $\log N = 5.124$
14. $\log N = -2.3145$
15. $\log x = 0.12483$
16. $\log y = -12.634$
17. $\ln N = 1.53$
18. $\ln N = 7.1542$
19. $\ln x = -12.0$
20. $\ln N = -1.00$
21. $\log_2 x = 2.18$
22. $\log_5 y = -1.72$

Do Exercises 23-38 as indicated:

23. Find x: $\log x - \log 23.4 = 1.563$
24. Find y: $0.065 = \dfrac{\ln y}{8.5}$

25. Find each of the following logarithms:
   (a) log 2.351 (b) log 23.51 (c) log 235.1
   (d) log 2351 (e) $\log 2.351 \times 10^9$
   What do you notice about all these logarithms?
   (The decimal part of a base 10 logarithm is called the mantissa of the logarithm and the whole number part is called the characteristic of the logarithm.)
26. Find each of the following logarithms:
   (a) log 24.2 (b) log 73.5 (c) log 99.9
   (d) log 11.1 (e) log 10.001
   What do you notice about all these logarithms? (See problem 25.)
27. Solve: $2^x = 9$.
28. Solve: $3.5^x = 7.6$
29. Solve: $e^x = 6.5234$
30. Solve: $3^{x+1} = 6.782$
31. In a solution, the hydrogen ion concentration is 0.00031. What is the pH?
32. The pH of a solution is 10.5. What is the hydrogen ion concentration?
33. If $d = \log\frac{I}{I_0}$, then find d if $I_0 = 95$ and $I = 163$.
34. If $d = \log\frac{I}{I_0}$, then find I if $I_0 = 95$ and $d = 2.4$.

◆ 35. *(Washington, Exercises 13-4, #29)* If the gain G (in decibels) of an electronic device is given by $G = 10\log\frac{P_0}{P_i}$ where P_0 is the output power and P_i is the input power in watts.
   (a) Determine the power gain if the input power is 0.785 W and the output power is 27.25 W.
   (b) Determine the output power if the gain is 4.8 decibels and the input power is 0.850 W.

◆ 36. *(Washington, Exercises 13-5, #46)* The intensity I of light decreases from its value I_0 as it passes a distance x through a medium. Given that $x = k(\ln I_0 - \ln I)$, where k is a constant depending on the medium. Let $k = 5.25$ cm.
   (a) Find x for $I = 0.850I_0$.
   (b) Find the ratio $\frac{I_0}{I}$ if $x = 0.70$ cm

37. If an amount of P dollars is invested at an annual interest rate r (expressed as a decimal), the value V of the investment after t years is given by $V = P(1+\frac{r}{n})^{nt}$, if interest is compounded n times a year. If $5000 is invested in an account earning

interest compounded 4 times per year, determine how long it will take the value to double the original investment if interest is calculated at a rate of

(a) 4.0%
(b) 6.0%
(c) 8.0%

38. The effect of inflation on the price, P, of a item may calculated using the formula $V = P(1+r)^n$ where r is the annual inflation rate (expressed as a decimal) and V is the price after n years.
(a) If the ticket at a certain movie house now costs $6.50 and the inflation rate is 3.5%, how long will it take the cost of the movie ticket to become $10.00?
(b) If the average price of a house is presently $115,000 and the inflation rate is 4.2%, how long will it take until the price of the house is $200,000?
(c) A child is given a piece of land valued at $25000 when it is born and later the land is worth $50,000 on the one of the child's birthdays. If the inflation rate is 4.0%, what birthday is the child celebrating?

Chapter 8

Systems of Non-linear Equations

8.1 Graphical Solutions to Non-linear Systems
(Washington, Section 14-1)

A non-linear equation is an equation that does not give a straight line when graphed. Such equations may involve either x^2 or y^2 terms or other functions such as trigonometric functions or logarithmic functions. Of primary interest are systems involving equations that can be written in the general form $Ax^2 + By^2 = C$ where A, B, and C represent real numbers. However, any system of equations, in which each equation can be solved for y in terms of x, can be solved by graphical means using the graphing calculator.

In order to solve a system of equations by graphical means, it is first necessary to solve each equation for y as a function of x. Equations of the form $Ax^2 + By^2 = C$ are not functions. However, it is possible to solve such an equation for y in terms of x and obtain two functions. Generally, one function will represent the top half of the graph and the other function will represent the bottom half of the graph.

Procedure G20. To graph equations involving y^2.

1. Solve the equation for y. There will be two functions: $y = f(x)$ and $y = g(x)$. In many cases it will be true that $g(x) = -f(x)$.
2. Store f(x) under one function name and g(x) for the second function name (see Procedure C10).
   (Example: $Y_1 = f(x)$
   $Y_2 = g(x)$)
   or
   If $g(x) = -f(x)$, we may do the following:
   (a) Store f(x) after a function name (Example: Y_1).
   (b) Move the cursor after the equal sign for a second function name (Example: Y_2).
   (c) Enter the negative sign: (–)
   (d) Obtain and enter the name of the first function.
   (Example: $Y_2 = -Y_1$) (See Procedure C11.)
3. Graph the functions.

After the two equations are graphed, the points of intersection can be found using Procedure G15.

Example 8.1: Solve the following system by graphing. Determine answers to two decimal places.

$$y = x^2 - 4$$
$$x^2 + 8y^2 = 64$$

Solution:

1. The first equation can be entered as is appears. However, the second equation must be solved for y in terms of x.

$$x^2 + 8y^2 = 64$$
$$8y^2 = 64 - x^2$$
$$y^2 = 8 - \frac{1}{8}x^2$$
$$\text{or } y = \pm\sqrt{8 - \frac{1}{8}x^2}$$

2. On the standard viewing rectangle, graph the second equation making use of Procedure G20 and graph the first equation in the form it is given. The graph is shown in Figure 8.1.
3. To find all the solutions to the system of equations it will be necessary to use Procedure G15 to find each of the four points of intersection. The points of intersection are:

(–2.58, 2.68), (2.58, 2.68), (1.09, –2.80), (–1.09,–2.80).

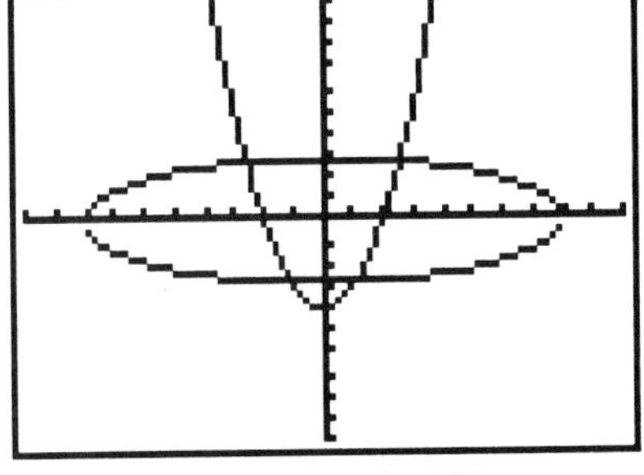

y:[–10, 10] x:[–10, 10]

Figure 8.1

The graphs in Example 8.1 were a parabola and an ellipse. Depending on the coefficients in the equations, a parabola and an ellipse may not intersect at all or may intersect at up to four points. This is often true if the system of equations involves x, y, x^2, and y^2.

In some cases it may be helpful to consider another method of finding the point of intersection of two graphs. For the system $y_1 = f(x)$ and $y_2 = g(x)$, consider the difference $y_1 - y_2 = f(x) - g(x)$. Let $F(x) = f(x) - g(x)$ and graph $y = F(x)$. A zero of the function F(x) will be the x-coordinate to a solution to the system $y_1 = f(x)$ and $y_2 = g(x)$ since at a point of intersection y_1 must equal y_2 and F(x) must be zero. Thus, another way to solve the system $y_1 = f(x)$ and $y_2 = g(x)$, is to graph $y = f(x) - g(x)$ and find the x-intercepts (the zeros of the function). The y-coordinates may be obtained by substituting the x-values into either f(x) or g(x).

Exercise 8.1

Solve each of the following systems of equations by graphing. Draw a sketch of the graphs obtained. Find answers to 3 significant digits. Identify any systems that do not have solutions.

1. $2x + 3y = 5$
   $y = x^2 - 8$

2. $5x - 2y = 7$
   $y = 6 - x^2$

3. $x^2 + y^2 = 25$
   $2x^2 - y = 11$

4. $3x^2 - 5y^2 = 24$
   $2x^2 + 2y^2 = 9$

5. $5x^2 - 12y^2 = 19$
   $x^2 - 4y = 15$

6. $x^2 - 3x + y = 7$
   $3x^2 = 38 - 2y^2$

7. $0.5x^2 + 0.2y^2 = 48$
   $5x + y = 6$

8. $3x^2 - 12x - y = 6$
   $6x + y = -21$

9. $xy = 8$
   $8x^2 - y^2 = 9$

10. $xy = 12$
    $3x + y = 12$

11. $y = \sin x$
    $x^2 - y = 1$

12. $y = \cos 2x$
    $y - x^3 = 2$

13. $y = \log x$
    $x^2 + y^2 = 2$

14. $y - 2^{x+3} = 5$
    $xy = 10$

The method of finding intersection points may also be used to solve a single equation. This is illustrated in Exercises 15 and 16.

15. Solve: $x^3 = 5x^2 + 6$.
    To do this let $y_1 = x^3$ and $y_2 = 5x^2 + 6$ and find the x-coordinates of the points of intersection.

16. Solve the equation $x^4 = 6 - x^2$ using the method of Exercise 15.

♦ 17. *(Washington, Exercises 14-1, #25)* A helicopter is located 5.358 miles northeast of a radio tower such that it is three times as far north as it is east of the tower. Find the northern and eastern components of the displacement from the tower.

♦ 18. *(Washington, Exercises 14-1, #26)* A 4.125 m insulation strip is placed completely around a rectangular solar panel with an area of 1.158 m^2. What are the dimensions of the panel?

19. An inflatable building has an equation $324.0x^2 + 1225y^2 = 396900$ (the top half). Cables to support the building against the wind are to be tied to one side. These cables can be considered as straight lines that go from the building to points on the positive x-axis and have equations $y = 15.0 - \dfrac{x}{4.25}$ and $y = 18.8 - \dfrac{x}{4.25}$. All measurements are given in feet. Find the points of intersection of the lines with the building. How long do these cables need to be?

20. The outer edge of one gear that is at an extreme point on a movable axis has an equation of $2.000x^2 + 5.120y^2 = 14.0625$ and that of a second gear has the equation $y^2 + x^2 - 11.300x = -25.920$. Measurements are in centimeters. Do the edges of the two gears intersect and if so at what point?

Chapter 9

Matrices

9.1 Matrices and Matrix Operations
(Washington, Sections 16-3 and 16-4)

A **matrix** can be defined as an ordered rectangular array of numbers. That is, it is a set of numbers arranged in rows and columns such that the order in which they are arranged is important. The numbers within a matrix are referred to as **elements** of the matrix. One way of classifying matrices is by size which depends on the number of rows and columns contained in the matrix. The **size** (or dimension) is given in the form **m x n** where m is the number of rows and n is the number of columns. Thus, a 2×3 matrix contains numbers arranged in two rows and three columns. A 4×1 matrix would be a matrix of four rows and one column. A matrix containing only one column is called a **column matrix**. A matrix containing one row and three columns is a 1×3 matrix. A matrix containing only one row is referred to as a **row matrix**.

In order to indicate that a rectangular array of numbers is a matrix, the array is enclosed either with large parentheses or with square brackets. The square brackets will be used here. Upper case letters are often used to name matrices -- such as matrix A or matrix B.

The elements of a matrix are identified according to the location of the element within the matrix by using a double subscript on a lower case letter. The first number in the subscript is the number of the row in which the element appears and the second number in the subscript is the number of the column in which the element appears. For example, a_{23} would represent the element in the second row, third column of matrix A and b_{41} would represent the element in the fourth row, first column of the matrix B. When using a calculator or a computer the double subscript is indicated by using two numbers (often within parentheses) separated by a comma. In this case a_{23} would appear as A(2,3) (or just as 2,3 when working with matrix A) and b_{41} would appear as B(4,1) (or as 4,1 when working with matrix B).

Example 9.1: For each of the following matrices, (a) give the size of the matrix, (b) identify the position occupied by the number 2 in each matrix and (c) give a_{32}, b_{31}, c_{12}, and d_{14}.

$$A = \begin{bmatrix} 5 & -4 & 1 \\ 4 & 4 & 2 \\ 1 & -1 & 0 \end{bmatrix}, \quad B = \begin{bmatrix} 2 \\ 4 \\ 3 \end{bmatrix}, \quad C = \begin{bmatrix} 1 & -2 \\ 2 & 3 \\ 0 & -1 \\ 5 & 4 \end{bmatrix}, \quad D = \begin{bmatrix} 1 & 0 & -3 & 2 \end{bmatrix}$$

Solution:

(a) The size of A is 3x3.
The size of B is 3x1.
The size of C is 4x2.
The size of D is 1x4.

(b) The number 2 occupies the position given by a_{23}, b_{11}, c_{21}, and d_{14}. In the notation that may be used on a calculator or computer these are represented as A(2,3), B(1,1), C(2,1), and D(1,4).

(c) $a_{32} = -1$, $b_{31} = 3$, $c_{12} = -2$, and $d_{14} = 2$.

Notice that, in Example 9.1, B is a column matrix and D is a row matrix. A matrix that contains the same number of rows as columns is referred to as a **square matrix**. In Example 9.1, matrix A is a square matrix.

Matrices are important in modern day mathematics for working a variety of problems. One application, the solving of systems of equations, is covered in Section 9.2. Graphing calculators have the ability of working with matrices and can be of great help in working matrix problems.

Procedure M1. To enter or modify a matrix.

1. Press the key MATRX. This gives the matrix menu.
2. Enter EDIT mode.
   On the TI-82:
   Highlight EDIT in the top row.
   On the TI-85:
   Select EDIT from the menu.
3. Select a name for the matrix.
   On the TI-82:
   Select one of the five matrices [A], [B], [C], [D], or [E]. This will be the new matrix or the matrix to be edited.
   On the TI-85:
   Enter the name for a new matrix or select the name of a matrix to be edited, then press ENTER. The name may be up to eight characters long (only 5 letters will show in name box).
4. The first line on the screen will contain the word MATRIX followed by the name of the matrix and two numbers separated by ×. These two numbers represent the size of the matrix. First enter the number

of rows and press ENTER, then the numbers of columns and press ENTER. To keep the values the same, use the cursor keys to move the cursor or just press ENTER.

5. Enter the elements of the matrix.

   On the TI-82:
   The element of the matrix to be entered or changed is highlighted and at the bottom of the screen is the row and column number of the element followed by its value.
   As a value is entered, the number at the bottom of the screen will change. To enter this value into the matrix, press ENTER. Values are entered by rows. Continue until all elements are correct.
   The cursor keys are used to move the highlight to elements to be changed.

   On the TI-85:
   On the left of the screen are the row and column numbers of elements of the matrix. A single column appears on the screen at a time. After each value is entered, press ENTER to go to the next value. Values are entered by row. (As each value is entered the calculator will jump to the next column.) The cursor keys may be used to go from one element to the next if an element is not to be changed. The menu selections ◁Col and Col▷ may be used to change columns.

6. Exit from matrix edit mode by pressing the key: QUIT. (Pressing CLEAR will clear the value for a given element.)

After entering a matrix, it is a good idea to view the matrix and to check to make sure it has been entered correctly.

Procedure M2. To select the name of a matrix and to view the matrix.

1. Press the key MATRIX. This gives the matrix menu.
2. Select the name of the matrix.

   On the TI-82:
   With NAMES highlighted in the top row, select one of the five matrices [A], [B], [C], [D], or [E].

   On the TI-85:
   Select NAMES from the matrix menu. From the secondary menu, select the name of the matrix.

3. View the matrix by pressing ENTER. In some cases the matrix may extend off the screen to the right or to the left. To see that part of the matrix not on the screen use the right or left cursor keys to scroll (move) the unseen part onto the screen.

In Example 9.2 we enter two square matrices into the calculator and check to make sure they are entered correctly.

Example 9.2: Enter the following matrices into the calculator as matrix M and matrix N.

$$M = \begin{bmatrix} 5 & 2 & 1 \\ 2 & 1 & -3 \\ 3 & -2 & 4 \end{bmatrix} \qquad N = \begin{bmatrix} 3 & -3 & 1 \\ -1 & 2 & 3 \\ 0 & -1 & -2 \end{bmatrix}$$

Solution:

1. Use Procedure M1 to enter matrix M into the calculator. Since M is a 3x3 matrix, enter the size as 3 rows and 3 columns.
2. After entering the size, enter the elements of the matrix.
3. Likewise, use Procedure M1 to enter matrix N.
4. Check to see if the matrices are entered correctly. Use Procedure M2 to view matrix M, then matrix N. If they have not been entered correctly, use Procedure M1 to make the necessary changes.

There are several operations that may be performed on matrices including the familiar operations of addition and subtraction. In order for matrices to be added or subtracted, the matrices must be of the same size. Division of matrices is not defined, but two kinds of multiplication do exist. One is **matrix multiplication** where two matrices are multiplied together and the other is **scalar multiplication** where a matrix is multiplied by a real number. We only consider the method of performing these operations on the calculator and not the techniques of doing the operations by hand.

Procedure M3. Addition, subtraction, scalar multiplication or multiplication of matrices.

See Procedure M2 for how to select the names of matrices.

1. Addition of two matrices:
   Select the name of the first matrix, press the addition sign, select the name of the second matrix, then press ENTER.
   (Example: [M] + [N])
2. Subtraction of two matrices:
   Select the name of the first matrix, press the subtraction sign, select the name of the second matrix, then press ENTER.
   (Example: [M] – [N])
3. Multiplication of matrix by a scalar:
   Enter the scalar, select the name of the matrix and press ENTER.
   (Example: 2.5 [M] or 2.5×[M])
4. Multiplication of two matrices:
   Select the name of the first matrix, select the name of the second matrix, and press ENTER.
   (Example: [M][N] or [M]×[N])

Many of these operations may be combined into one statement. The result of a matrix calculation is stored under ANS.

We now perform the operations of addition, subtraction, and scalar multiplication on the graphing calculator.

Example 9.3: For the matrices M and N given in Example 9.2, find M + N, M – N, 3M, and 3M – 2N.

Solution:

1. Use Procedure M3 to add M and N.

$$M + N = \begin{bmatrix} 8 & -1 & 2 \\ 1 & 3 & 0 \\ 3 & -3 & 2 \end{bmatrix}$$

2. Use Procedure M3 to subtract N form M.

$$M - N = \begin{bmatrix} 2 & 5 & 0 \\ 3 & -1 & -6 \\ 3 & -1 & 6 \end{bmatrix}$$

3. Use Procedure M3 to multiply M by the scalar 3.

$$3M = \begin{bmatrix} 15 & 6 & 3 \\ 6 & 3 & -9 \\ 9 & -6 & 12 \end{bmatrix}$$

4. Use Procedure M3 to multiply M by 3 then subtract 2 times N.

$$3M - 2N = \begin{bmatrix} 5 & 12 & 1 \\ 8 & -1 & -15 \\ 9 & -4 & 16 \end{bmatrix}$$

Addition, subtraction, and scalar multiplication of simple matrices are easily done by hand. However, matrix multiplication is not as readily handled by hand. **Matrix multiplication is only defined when the number of columns in the first matrix equals the number of rows in the second matrix.**

Example 9.4: Use matrices M and N of Example 9.2 to find the matrix product MN and then the product NM.

Solution: 1. Use Procedure M3 to find the product of M and N.

$$MN = \begin{bmatrix} 13 & -12 & 9 \\ 5 & -1 & 11 \\ 11 & -17 & -11 \end{bmatrix}$$

2. Use Procedure M3 to find the product of N and M.

$$NM = \begin{bmatrix} 12 & 1 & 16 \\ 8 & -6 & 5 \\ -8 & 3 & -5 \end{bmatrix}$$

In Example 9.4, we see that MN does not equal NM. In general, the commutative law of algebra (ab = ba) does not hold for matrix multiplication.

Exercise 9.1

In Exercises 1-22, use the matrices.

$$A = \begin{bmatrix} 1 & 2 & -4 \\ 3 & 2 & -1 \\ 5 & -2 & 7 \end{bmatrix}, \quad B = \begin{bmatrix} 4 \\ 7 \\ 8 \end{bmatrix}, \quad C = \begin{bmatrix} 6 & -7 & -1 \\ 3 & 4 & -5 \\ 3 & -5 & 7 \end{bmatrix},$$

$$D = \begin{bmatrix} -3 & 3 & -1 & 7 \end{bmatrix}, \quad E = \begin{bmatrix} 1 & 2 & 3 & -5 \end{bmatrix}$$

1. Give the size of each matrix A, B, C, and D.
2. Which of the given matrices are square matrices?
3. Identify the position occupied by the number –1 in each matrix.
4. Identify the position occupied by the number 7 in each matrix.
5. List all pairs of these matrices that may be added or subtracted?
6. Give all pairs, in correct order, of these matrices that can be multiplied.

In Exercises 7-22, enter the given matrices into your calculator and use the calculator to perform the indicated operation. If it is not possible to perform the operation, indicate why not.

7. A + C
8. D – E
9. D – 2B
10. 3A
11. 4A – 2C
12. 3D – 7E
13. –3C – 7A
14. 4E – C
15. AB
16. BA
17. AC
18. CA
19. A(CB)
20. B + CB
21. AB – B
22. AC – C

♦ 23. *(Washington, Exercises 16-4, #31)* Using matrices A and C given at the beginning of this exercise set, show that $A^2 - C^2 = (A + C)(A - C)$.

♦ 24. *(Washington, Exercises 16-4, #34)* In analyzing the motion of a robotics mechanism, the following matrix multiplication is used. Perform the multiplication.

$$\begin{bmatrix} \cos 35.7° & -\sin 35.7° & 0 \\ \sin 35.7° & \cos 35.7° & 0 \\ 0 & 0 & 1 \end{bmatrix} \begin{bmatrix} 2.5 \\ 4.2 \\ 0 \end{bmatrix}$$

9.2 The Inverse Matrix and Solving a System of Equations *(Washington, Section 16-5 and 16-6)*

The **identity** matrix is a square matrix with one's as elements in the positions where the number of the row and the number of the column are the same and zeros elsewhere. The identity matrix is denoted by the letter I. The identity matrix has the property such that AI = A and IA = A for any matrix A. When I appears in a matrix expression, it is assumed to have a size that will make the operations of the expression valid. The 2x2 and 3x3 identity matrices are

$$I = \begin{bmatrix} 1 & 0 \\ 0 & 1 \end{bmatrix}, \quad I = \begin{bmatrix} 1 & 0 & 0 \\ 0 & 1 & 0 \\ 0 & 0 & 1 \end{bmatrix}$$

A square matrix A has an **inverse** matrix B only if AB = I and BA = I. The notation A^{-1} is used to denote the inverse of matrix A. (Note that the −1 as used here does not mean the reciprocal.)

Two other operations on matrices are of interest. One is that of finding the determinant of a matrix and the other is finding the transpose of a matrix. Every square matrix has associated with it a number called the **determinant**. A method of solving systems of equations called Cramer's rule makes use of determinants. We will not consider Cramer's rule, but will consider the use of determinants to determine whether or not a matrix has an inverse. If the determinant of a square matrix is not zero then the matrix has an inverse. Such a matrix is said to be **non-singular**. If the determinant is zero, then the matrix does not have an inverse and the matrix is said to be **singular**.

The **transpose** of a matrix is the matrix formed by interchanging the rows and columns of a matrix. The transpose of matrix A is indicated by the notation A^T.

$$\text{If } A = \begin{bmatrix} 1 & -2 & 3 \\ 2 & 3 & -1 \\ 0 & -1 & -3 \end{bmatrix} \text{ then } A^T = \begin{bmatrix} 1 & 2 & 0 \\ -2 & 3 & -1 \\ 3 & -1 & -3 \end{bmatrix}$$

Procedure M4. To find the determinant, inverse, and transpose of a matrix.

1 To find the determinant:

On the TI-82:
Press the key MATRIX
Highlight MATH in top row.
From the menu, select det.

On the TI-85:
From the matrix menu, select MATH.

Then select the name of the matrix (Procedure M2) and press ENTER.
(Example: det[A] or det A}

2. To find the inverse:
Select the name of the matrix.
Press the key x^{-1} and press ENTER.
(Example: $[A]^{-1}$ or A^{-1})

3. To find the transpose:
Select the name of the matrix, then

On the TI-82:
Highlight MATH in the top row.

On the TI-85:
From the matrix menu, select MATH.

From the menu, select ^{T} and press ENTER.
(Example: $[A]^T$ or A^T)

Sometimes it is desirable to store one matrix under another name or to store the results of a calculation. Care needs to be taken since whenever something is stored under a variable name, it erases whatever was previously stored under that name.

Procedure M5. To store a matrix.

1. Select the name of the first matrix or enter a matrix expression.
2. Press the key STO▷
3. Select the name of the second matrix and press ENTER.
   (Example: [A] + [B] → [C] or A + B →C)
   On the TI-85, we may also store a matrix by using the equal sign: C = A + B.

Note: Storing a matrix for a second matrix, erases the previous contents of the second matrix.

Finding the transpose of a matrix is easily done by hand. However, finding the inverse or determinant of a matrix by hand can become rather involved. With the graphing calculator, it is a relatively easy process.

Example 9.5: Find the inverse of matrix A, store this as matrix B, then show that AB = I and BA = I. Also, find the transpose of B. Round numbers to three decimal places.

$$A = \begin{bmatrix} 1 & -2 & 3 \\ 2 & 3 & -1 \\ 0 & -1 & -3 \end{bmatrix}$$

Solution:

1. Enter the given matrix into the calculator as matrix A. (See procedure M1.)
2. Since the inverse matrix may be in decimal form and, thus, may be carried out to 10 decimal places, change the mode so that the number of decimal places displayed by the calculator is three. (See Procedure C6)
3. Using Procedure M4 to find the inverse of matrix A we find.

$$A^{-1} = \begin{bmatrix} .357 & .321 & .250 \\ -.214 & .107 & -.250 \\ .071 & -.036 & -.250 \end{bmatrix}$$

4. Use Procedure M5 to store the inverse of A as matrix B.

5. Find the product of matrix A and matrix B (AB) and the product of matrix B and matrix A. (BA). In both cases, a matrix similar to the identity matrix appears except that it may contain numbers such as 1.000E–13 or 2.000E–14. These numbers with large negative exponents are essentially zero and may be taken as zero. The matrix obtained in each case is the identity matrix I and B is the inverse of A.
6. Using Procedure M4 to find the transpose of matrix B, we have

$$B^T = \begin{bmatrix} .357 & -.214 & .071 \\ .321 & .107 & -.036 \\ .250 & -.250 & -.250 \end{bmatrix}$$

Since matrix multiplication does not satisfy the commutative property, to show that A is the inverse of B, it is necessary to show both that AB = I and that BA = I.

An important use of an inverse matrix is to solve a system of equations. To solve a system of equations we first determine the coefficient matrix after writing the system of equations in standard form. **Standard form** for a system of equations means writing the equations so that all the variables are in the same order to the left of the equal sign and the constants to each equation are on the right side of the equal sign. For the following system of equations

$$\begin{aligned} &2x + 3y = 4 + z \\ &x + 4z - 8 = 2y \\ &3x + 2y = 12 + 3z \end{aligned}$$

the standard form is

$$\begin{aligned} &2x + 3y - z = 4 \\ &x - 2y + 4z = 8 \\ &3x + 2y - 3z = 12 \end{aligned}$$

The **coefficient matrix** of a system of equations is the matrix formed by the coefficients of the unknowns when the equations are in standard form. The **constant matrix** is the column matrix formed by the constants when the equations are in standard form. The **unknown matrix** is the column matrix formed by the variables in the system of equations. For the given system, the coefficient matrix, A, the unknown matrix, X, and the constant matrix, C are

$$A = \begin{bmatrix} 2 & 3 & -1 \\ 1 & -2 & 4 \\ 3 & 2 & -3 \end{bmatrix}, \quad X = \begin{bmatrix} x \\ y \\ z \end{bmatrix}, \quad C = \begin{bmatrix} 4 \\ 8 \\ 12 \end{bmatrix}$$

If A and X are multiplied together, the result is the left side of the system of equations

$$AX = \begin{bmatrix} 2 & 3 & -1 \\ 1 & -2 & 4 \\ 3 & 2 & -3 \end{bmatrix} \begin{bmatrix} X \\ Y \\ Z \end{bmatrix} = \begin{bmatrix} 2X+3Y-Z \\ X-2Y+4Z \\ 3X+2Y-3Z \end{bmatrix}$$

For a system of equations this last matrix equals the constant matrix C. Therefore, the given system of equations, and in general, any system, may be written as a matrix equation in the form

$$A\,X = C$$

To solve the matrix equation, multiply both sides by the matrix A^{-1}.

$$A^{-1}A\,X = A^{-1}C$$

or $\quad I\,X = A^{-1}C \qquad$ [since $A^{-1}A = I$]

or $\quad X = A^{-1}C \qquad$ [since $I\,X = X$]

This indicates that to find the unknown matrix X and, thus, to solve this system of equations, we need to find the inverse of A and multiply it by the coefficient matrix C.

Example 9.6: Use the inverse of the coefficient matrix to find the solutions to the following system of equations, then verify that the matrix found is the solution matrix. Round values to four decimal places.

$$2x + 3y - z = 4$$
$$x - 2y + 4z = 8$$
$$3x + 2y - 3z = 12$$

Solution:

1. Enter the coefficient matrix into the calculator as the 3x3 matrix A.
2. Enter the constant matrix into the calculator as the 3x1 matrix C.
3. Since $X = A^{-1}C$, multiply A^{-1} by C (see Procedure M4).
   From the results, the solution to the system of equations, rounded to four decimal places is given by

$$X = \begin{bmatrix} x \\ y \\ z \end{bmatrix} = \begin{bmatrix} 5.0909 \\ -2.1818 \\ -.3636 \end{bmatrix}$$

By equality of matrices this means that x = 5.0909, y = –2.1818, and z = – .3636.

4. To verify the results, store the product $A^{-1}C$ as matrix B. (See Procedure M5)
5. Next, multiply matrices A and B. If the solution is correct, the result should be equal to matrix C.

The process illustrated in Example 9.6 turns out to be quite powerful. It gives a relatively simple method to solve any system of equations if the system of equations has a unique solution. Since a system of equations may not always have a solution or may have infinite many solutions, it is wise to determine if a solution exists before attempting the matrix calculation. We do this by evaluating the determinant of the coefficient matrix. **If the determinant of the coefficient matrix is zero, then the coefficient matrix is a singular matrix and does not have an inverse and the system of equations does not have a unique solution.** Such a system may have either no solutions or an infinite number of solutions. When this occurs, other methods, such as the Gauss-Jordan Method, need to be employed to determine if there are no solutions or an infinite number of solutions. If there are an infinite number of solutions, the form of these solutions may be obtained from the Gauss-Jordan Method.

Example 9.7: Find the determinant of the coefficient matrices for the following system of equations and determine if each system has a unique solution.

(a) $2x + 4y = 3$
$5x - 3y = 8$

(b) $2x - 3y + z = 2$
$x + 2y - 2z = 7$
$3x - 8y + 4z = 9$

Solution: 1. The coefficient matrices are

$$\text{(a)}\ A = \begin{bmatrix} 2 & 4 \\ 5 & -3 \end{bmatrix} \quad \text{and} \quad \text{(b)}\ B = \begin{bmatrix} 2 & -3 & 1 \\ 1 & 2 & -2 \\ 3 & -8 & 4 \end{bmatrix}$$

2. Enter these matrices into the calculator as matrices A and B.

3. Use Procedure M4 to find the determinants of each of these matrices. The determinant of matrix A is –26. The determinant of matrix B is –1.4E–11 or -1.4×10^{-11}. The value -1.4×10^{-11} is essentially zero and we take the determinant of matrix B to be zero.
4. Since the determinant of matrix A is not zero, system (a) has a unique solution and since the determinant of matrix B is zero, system (b) does not have a unique solution.

Exercise 9.2

In Exercises 1-8, first determine (a) the transpose of the given matrix, and (b) the determinant of the given matrix. If the determinant is not zero, find the inverse of the matrix and prove by multiplication that the inverse is really the inverse. Round all values to 3 decimal places.

1. $\begin{bmatrix} 2 & -5 \\ -3 & 4 \end{bmatrix}$

2. $\begin{bmatrix} -1 & 4 \\ -3 & 12 \end{bmatrix}$

3. $\begin{bmatrix} 3 & 2 & -1 \\ -4 & 0 & 3 \\ 5 & 3 & -2 \end{bmatrix}$

4. $\begin{bmatrix} -6 & 12 & 7 \\ 10 & 5 & -4 \\ -3 & 8 & 11 \end{bmatrix}$

5. $\begin{bmatrix} 1 & -2 & -4 \\ -5 & 3 & 1 \\ 7 & -7 & -9 \end{bmatrix}$

6. $\begin{bmatrix} 10 & 7 & 8 \\ 3 & 2 & -5 \\ -7 & -8 & 15 \end{bmatrix}$

7. $\begin{bmatrix} 0.03 & 0.12 & 0.10 \\ -0.04 & 0.45 & 0.32 \\ 0.11 & 0.95 & 0.04 \end{bmatrix}$

8. $\begin{bmatrix} -0.03 & 0.15 & 0.12 \\ 0.15 & 0.12 & 0.05 \\ -0.24 & 0.33 & 0.31 \end{bmatrix}$

In Exercises 9-20, (a) find the determinant of the coefficient matrix, (b) find the inverse matrix, and (c) use the inverse matrix method to solve the given system of equations.

9. $3x + 2y = 5$
   $2x - 3y = 7$

10. $5x - 3y = 8$
    $-2x + 4y = 3$

11. $6x - 2y = 5$
    $-9x + 3y = 2$

12. $0.05x + 0.25y = 2.35$
    $0.10x - 0.50y = 7.85$

13. $2x - 3y + 6z = 7$
$-3x + 2y - 3z = 12$
$5x + 7y + 3z = 9$

14. $7x - 8y + 6z = -12$
$4x + 2y - 7z = 9$
$-5x + 7y + 6z = 15$

15. $3x - 2y = z - 7$
$2x - 3z = 4 + y$
$-x = 7 - 2y + 3z$

16. $3x - 4y = 5z - 3$
$x = -7y - 2z - 7$
$6y - 6z = 12 - 8x$

17. $0.21x - 0.25y + 0.12z = 265$
$0.17x + 0.33y - 0.20z = 125$
$0.11x - 0.45y + 0.16z = 225$

18. $x = 6y$
$x + y + z = 16$
$0.04x + 0.07y = 0.25$

19. $2x = 3y$
$0.10x + 0.15y + 0.25z = 0$
$5.4x + 8.4y = 1 + 6.2z$

20. $i_1 + i_2 + i_3 = 0$
$12i_1 - 8.5i_2 = 5.15$
$8.5i_1 - 15i_3 = -2.14$

21. The motion of the rotation of a point through an angle of 37.5° about the origin on a computer screen is given by matrix R. Find R^{-1} that describes the inverse rotation.

$$R = \begin{bmatrix} \cos 37.5° & -\sin 37.5° & 0.0000 \\ \sin 37.5° & \cos 37.5° & 0.0000 \\ 0.0000 & 0.0000 & 1.000 \end{bmatrix}$$

♦ 22. *(Washington, Exercises 16-5, #32)* The rotations of a robot arm is represented by matrix R. The values represent trigonometric functions of the angle of rotation. Find R^{-1}.

$$R = \begin{bmatrix} .5736 & 0.0000 & -0.8192 \\ 0.0000 & 1.0000 & 0.0000 \\ 0.8192 & 0.0000 & 0.5736 \end{bmatrix}$$

Solve Exercises 23 and 24 by using the inverse matrix of the coefficient matrix.

23. Three different alloys are made up of zinc, lead, and copper. Alloy A contains 55.0% zinc, 32.0% lead, and 13.0% copper. Alloy B contains 38.0% zinc, 32.0% lead, and 30.0% copper. Alloy C contains 34.5% zinc and 65.5% lead. It is desired to mix these three alloys to obtain 200.0 grams of a mixture that contains 45.0% zinc, 37.5% lead, and 17.5% copper. To determine how many grams of each alloy are needed, it is necessary to solve the following system of equations

where we let x = number of grams of alloy A, y = number of grams of alloy B, and z = number of grams of alloy C.

$$\begin{aligned} 0.550x + 0.380y + 0.345z &= 0.450(200) \\ 0.320x + 0.320y + 0.655z &= 0.375(200) \\ 0.130x + 0.300y \qquad\qquad &= 0.175(200) \end{aligned}$$

♦ 24. *(Washington, Exercises 16-6, #22)* In applying Kirchhoff's laws to an electrical circuit the following equations are found. $I_A, I_B, \text{and } I_C$ are three currents, in amperes, in the circuit. Find $I_A, I_B, \text{and } I_C$.

$$\begin{aligned} I_A + I_B + I_C &= 0 \\ 2.35I_A - 4.75I_B &= 6.34 \\ 4.75I_B - 1.20I_C &= -3.24 \end{aligned}$$

9.3 Row Operations on Matrices
(Washington, Sections 16-5 and S-1)

Since it not always possible to use the inverse matrix method to solve a system of equations, it is sometimes helpful to use another method called the Gauss-Jordan method. The Gauss-Jordan method may be used to solve any system of linear equations, but it is particularly useful in finding the form of a solution when a system of equations has an infinite number of solutions or in determining if there are no solutions to a system of equations. The Gauss-Jordan method makes use of what are called row operations on matrices.

Row operations may also be used to find the inverse of a matrix. However, we will not use this method to find the inverse, since on a graphing calculator, the inverse matrix may be found directly.

There are three row operations, which when performed on matrices, yield equivalent matrices (not equal matrices). Many graphing calculators have a provision for performing the matrix row operations.

Matrix Row Operations

1. Any two rows may be interchanged.
2. Any row may be multiplied (or divided) by a non-zero constant.
3. Any row may be multiplied by a non-zero constant and added to a second row, replacing the second row.

Procedure M6: Row operations on a matrix.

1. Obtain the matrix operations menu by first pressing the key MATRX, then:

On the TI-82:	On the TI-85:
Highlight MATH in the top row.	From the matrix menu, select OPS (then press MORE).

2. Select the desired operation from the menu. Names of matrices are selected as in Procedure M1. (The left hand column gives the selection, form, and example for the TI-82 and the right hand column given the selection, form, and example for the TI-85.)
   (a) To interchange two rows on a matrix, select

TI-82	TI-85
rowSwap(	rSwap
Form:	Form:
rowSwap(name, row1, row2)	rSwap(name, row1, row2)
Example:	Example:
rowSwap([A], 1, 3)	rSwap(A, 1, 3)

   (Interchanges rows 1 and 3 of matrix A.)

   (b) To multiply a row by a constant, select:

TI-82	TI-85
*row(	multR
Form:	Form:
*row(multiplier, name, row)	multR(multiplier, name, row)
Example:	Example:
*row(1/2, [A], 3)	multR(1/2, A, 3)

   (Multiplies row 3 of matrix A by ½.)
   To divide a row by a constant, multiply by the reciprocal.

   (c) To multiply a row by a constant and add to another row, select

TI-82	TI-85
*row+(	mRAdd
Form:	Form:
*row+(mult,name,row1,row2)	mRAdd(mult,name,row1,row2)
Example:	Example:
*row+(−5,[A], 1, 3)	mRAdd(−5, A, 1, 3)

   (Multiplies row 1 of matrix A by −5 and adds the result to row 3 and replaces row 3.)

3. Press ENTER.

To solve a system of system of equations by the Gauss-Jordan method, the system of equations is first written in standard form. As we have previously seen the matrix form of a system of equations is AX = C. For the Gauss-Jordan method, we form an **augmented matrix** of the form (A|C) where A is the coefficient matrix and C is the constant matrix. We then perform matrix row operations on this matrix as demonstrated in Example 9.8. The goal is to get this matrix into the form (I|Y), where I is the identity matrix. If this can be accomplished, then Y is the solution matrix.

Example 9.8: Solve the system of following system of equations by the Gauss-Jordan method.

$$x + 2y = 5$$
$$3x + 8y + z = 5$$
$$2x - z = 12$$

Solution:

1. Since these equations are already in standard form, we write the augmented matrix for this system

$$\left[\begin{array}{ccc|c} 1 & 2 & 0 & 5 \\ 3 & 8 & 1 & 5 \\ 2 & 0 & -1 & 12 \end{array}\right]$$

2. First, make sure element $a_{11} = 1$. In this case, this is already true. If it were not, we would perform row operation 2 to make it a 1. Next we want to make a_{21} and a_{31} both zero. We do this by performing row operation 3. Use Procedure M6. 2(c) to multiply row 1 by –3 and add to row 2 and to multiply row 1 by –2 and add to row 3. The result is

$$\left[\begin{array}{ccc|c} 1 & 2 & 0 & 5 \\ 0 & 2 & 1 & -10 \\ 0 & -4 & -1 & 2 \end{array}\right]$$

3. Next, we make $a_{22} = 1$ by multiplying row 2 by $\frac{1}{2}$ using row operation 2. Use Procedure M6. 2(b).

$$\left[\begin{array}{ccc|c} 1 & 2 & 0 & 5 \\ 0 & 1 & .5 & -5 \\ 0 & -4 & -1 & 2 \end{array}\right]$$

4. Then, we will make a_{12} and a_{32} both zero by using row operation 3. Use Procedure M6. 2(c) to multiply row 2 by –2 and add to row 1 and multiply row 2 by 4 and add to row 3.

$$\left[\begin{array}{ccc|c} 1 & 0 & -1 & 15 \\ 0 & 1 & .5 & -5 \\ 0 & 0 & 1 & -18 \end{array}\right]$$

5. Next we make sure $a_{33} = 1$. If not, use row operation 2 to make it a one. Then make both a_{13} and a_{23} zero by using row operation 3. Use Procedure M6. 2(c) to multiply row 3 by –.5 and add to row 2 and to multiply row 3 by 1 and add to row 1.

$$\left[\begin{array}{ccc|c} 1 & 0 & 0 & -3 \\ 0 & 1 & 0 & 4 \\ 0 & 0 & 1 & -18 \end{array}\right]$$

6. We now have achieved the correct form. When we write this last matrix as a system of equations we obtain the solution to the given system of equations

$$x = -3, \; y = 4, \; \text{and } z = -18.$$

If, when performing row operations on an augmented matrix of the form (A|C), a row of all zeros is obtained, then the system of equations has an infinite number of solutions. If a row is obtained that contains all zeros except for the last element of that row (to the right of the vertical bar), then such a system has no solutions.

Example 9.9: Solve the following system of equations using the Gauss-Jordan method.

$$\begin{aligned} x + 2y - 3z &= 5 \\ 2x + y + 2z &= 4 \\ x + 5y - 11z &= 11 \end{aligned}$$

Solution:

1. These equations are already in standard form. Thus, write the augmented matrix for this system.

$$\left[\begin{array}{ccc|c} 1 & 2 & -3 & 5 \\ 2 & 1 & 2 & 4 \\ 1 & 5 & -11 & 11 \end{array}\right]$$

2. Since element $a_{11} = 1$ we proceed to make a_{21} and a_{31} both zero. We do this by performing row operation 3. Use Procedure M6. 2(c) to multiply row 1 by –2 and add to row 2 and to multiply row 1 by –1 and add to row 3. The result is

$$\left[\begin{array}{ccc|c} 1 & 2 & -3 & 5 \\ 0 & -3 & 8 & -6 \\ 0 & 3 & -8 & 6 \end{array}\right]$$

3. Next we will make $a_{22} = 1$ by multiplying row 2 by $-\frac{1}{3}$ using row operation 2. Use Procedure M6. 2(b). The result, rounded to three decimal places is:

$$\left[\begin{array}{ccc|c} 1 & 2 & -3 & 5 \\ 0 & 1 & -2.667 & 2 \\ 0 & 3 & -8 & 6 \end{array}\right]$$

4. Next we will make a_{12} and a_{32} both zero by using row operation 3. Use Procedure M6. 2(c) to multiply row 2 by −2 and add to row 1 and multiply row 2 by −3 and add to row 3.

$$\left[\begin{array}{ccc|c} 1 & 0 & 2.333 & 1 \\ 0 & 1 & -2.667 & 2 \\ 0 & 0 & -2E-13 & 0 \end{array}\right]$$

5. The quantity −2E−13 in the last row is -2×10^{-13} or, for most practical purposes, zero. Thus, the last row is all zeros and this system has an infinite number of solutions. The equations given by the first two rows are

$$x+\frac{7}{3}z=1 \qquad\qquad x=1-\frac{7}{3}z$$
$$\text{or}$$
$$y-\frac{8}{3}z=2 \qquad\qquad y=2+\frac{8}{3}z$$

The solutions are the set of numbers (x, y, z) such that if we know any one of the three variables, we determine the other two variables from these equations. For example, to obtain one of the solutions let $z = 3$. Then, $x = -6$ and $y = 10$. Another solution may be obtained by letting $z = 0$. Then, $x = 1$ and $y = 2$.

Exercise 9.3

Solve each of the following systems of equations by setting up the augmented matrix and using row operations. Identify those that have no solutions and those that have an infinite number of solutions.

1. $8x - 5y = 25.5$
   $5x = 32.7 - 2y$

2. $3.52x + 6.24y = 125.45$
   $y = 6.54x - 54.60$

3. $3x + 4y = 64.8$
   $8y = 25.5 - 6x$

4. $0.22x + 0.35y = 2.45$
   $0.64x - 0.65y = 3.47$

5. $x + 2y - z = -1$
   $-5y - 3x + z = -7$
   $2x + 2z - y = 1$

6. $2x - 3y + 5z = 13$
   $-y + z - 3x = 9$
   $3x + 16z - 10y = 48$

7. $6x - 3y + 2z = 29$
   $x + 2y - 3z = 7$
   $3x - 9y + 11z = 8$

8. $z = x - 2y$
   $0.3x + 0.4y - 0.5z = 1.5$
   $1.1x - 5.2y = 0.5z = 0$

9. $x + y + z = 15$
   $2.5x + 3.0y - 1.5z = 7$
   $6.0x + 7.0y - 2z = 29$

10. $25x + 30y - 15z = 125$
    $10x - 15y + 12z = 173$
    $5x - 34z = 221$

11. $3.5x + 2.1y + 2.8z = 55$
    $1.4x - 1.3y - 2.3z = 17$
    $0.7x + 5.7y + 7.4z = 21$

12. $3x + 5y + 4z + w = 7$
    $2x - 3y - 6z + w = 9$
    $x + 4y + 2z - w = 5$
    $4x - 2y - 4z = 6$

♦ 13. *(Washington, Exercises S-1, #23)* Three machines together produce 672 parts each hour. Twice the production of the second machine is 12 parts more than the sum of the other two machines. If the first operates for 3.5 hours and the others operate for 2.5 hours, 1992 parts are produced. Find the production rate of each machine.

♦ 14. *(Washington, Exercises S-1, #24)* A total of \$12,550 is invested, part at 6.5%, part at 6.0%, and part at 5.5%, yielding a total annual interest of \$773.20. The income from the 6.5% part yields \$35.40 more than that for the other two parts combined. How much is invested at each rate?

Chapter 10

Graphing of Inequalities

10.1 Graphing of Inequalities
(Washington, Sections 17-5)

In this section we are interested in solving inequalities in two variables by graphing. Inequalities contain two algebraic expressions separated by one of the inequality signs ($>$, $<$, $\leq$, or $\geq$). A **solution to an inequality** in two variables consists of all pairs of numbers (x, y) that satisfy the inequality. The pairs of numbers (0, 1), (1, 0), (−2, 2) and (1, −1) are all solutions of the inequality $3x - 2y \leq 5$ since when the values are substituted for x and y the inequality is a true statement. Inequalities generally have an infinite number of solutions. To visualize the solutions to an inequality, we graph the inequality and shade in the region corresponding to all pairs of numbers that are solutions.

Before graphing an inequality, it is first necessary to solve the inequality for y as we did when graphing equations. Rules for working with inequalities are similar to, but not the same as, those for working with equations. A major difference occurs when both sides of an inequality are multiplied or divided by the same negative value.

Rules for Inequalities

1. The same term may be added or subtracted to both sides of the inequality without changing the direction of the inequality sign.
2. Both sides of an inequality may be multiplied or divided by the same positive value without changing the direction of the inequality sign.
3. If both sides of an inequality are multiplied or divided by the same negative value, then the direction of the inequality sign is reversed.

To graph an inequality on a graphing calculator, first graph the equation corresponding to the inequality and then shade the region that indicates the solution to the inequality by using the shade function of the calculator. When using the shade function it is sometimes desirable to use special variables such as Ymin, Ymax, Y_1, etc. Procedure G21 indicates how to obtain these special values and Procedure G22 indicates how to use the shade function.

Procedure G21. To obtain special Y-variables.

On the TI-82:

1. To obtain Ymin or Ymax
   Press the key: VARS
   From the menu,
   select: Window..
   Select Ymin or Ymax
2. To obtain variable names Y_1, Y_2, etc.
   Press the key: Y-VARS
   From the menu, select:
   Function
   Select the function name.

On the TI-85:

The best way is perhaps to just enter the name from the keyboard. When entering the name from the keyboard, be sure the correct upper or lower case is entered.

or

1. Press the key VARS
2. Select ALL
3. Press F1 to page down till desired variable is on screen.
4. Use cursor keys to select desired variable.
5. Press ENTER

Procedure G22. To shade a region of a graph.

1. Enter the desired function for a function name. Make sure all functions not wanted are deleted or turned off.
2. Select appropriate window values and graph the function.
3. Select the Shade command:

On the TI-82:
Press the key: DRAW
From the menu, select 7:Shade(
Form is
Shade(L, U, D, Lt, Rt)

On the TI-85:
From the GRAPH menu, select DRAW (after pressing MORE) and select Shade
Form is Shade(L, U, Lt, Rt)

Where:
L is lower boundary (value or function) of the shaded area.
U is upper boundary (value or function) of the shaded area.
D is the density value (a digit from 1 to 8) that determines the spacing of the vertical lines the calculator draws to shade the region. The higher the value for the density, the more widely spaced the lines. If the density value is omitted, the shading is solid.
Lt is the left most x-value.
Rt is the right most x-value.
[D, Lt, Rt are optional - but on the TI-82, if Lt and Rt at included, D must be also included.]
(Example: Shade(−10,Y_1) shades in the area above $y = -10$ and below the graph of Y_1)

4. Enter the appropriate values for L, U, and/or D and left and right x-values. Then, press ENTER.

   Note: To shade above the function represented by Y_1, use Shade(Y_1, Ymax).
   To shade below the function represented by Y_1, use Shade(Ymin, Y_1).

When graphing inequalities involving $\leq$ or $\geq$, the curve is included as part of the solution, but when graphing inequalities involving $>$ or $<$, the curve itself is not included as part of the solution. On the graphing calculator it is not possible to show the difference between these cases. However, when sketching the graph by hand, use a solid curve for the graph of the equation for inequalities involving $\leq$ or $\geq$, and a dashed curve for the graph of the equation for the inequalities involving $>$ or $<$. After the curve is drawn, shade in the proper area.

Example 10.1: Graph the solution to the inequality

$$x - 2y > 6.$$

Solution:

1. First, solve the inequality for y.

$$x - 2y > 6$$
$$-2y > 6 - x$$
$$y < -3 + \frac{1}{2}x \quad \text{[divide both sides by } -2\text{, change direction of inequality]}$$

2. Graph the equation $y = -3 + \frac{1}{2}x$ by entering $-3 + \frac{1}{2}x$ for Y_1. Y_1 should be the only function turned on. Select the standard viewing rectangle.
3. Next shade in the area corresponding to the solution. Since the inequality is $y < -3 + \frac{1}{2}x$, we want the area such that y is less than the line $y = -3 + \frac{1}{2}x$
4. Use Procedure G22 to shade the region below Y_1. For the lower boundary of the shaded area, we select Ymin. For the upper boundary of the shaded area, we want the graph of

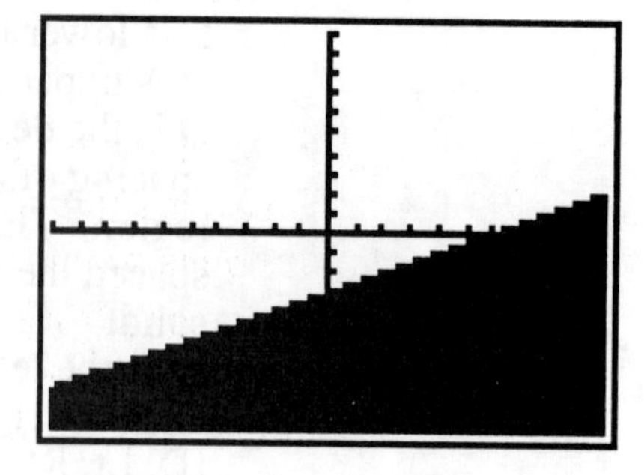

Figure 10.1

$y = -3 + \frac{1}{2}x$ or Y_1. (Procedure G21).

5. The statement needed is Shade(Ymin,Y_1). Press ENTER. The graph is in Figure 10.1.

Any pair of numbers x and y that correspond to points in the shaded region will be solutions to the given inequality. As a check some selected points may be substituted into the inequality to determine if the inequality is true at these points. Remember when sketching this graph on a sheet of paper, to make the line $y = -3 + \frac{1}{2}x$ a dashed line.

In certain applications, it is helpful to find solutions to systems of inequalities. A system of inequalities is a set of two or more inequalities with two or more variables. A **solution to a system of linear inequalities** in two variables x and y consists of all pairs of numbers (x, y) that satisfy all the inequalities in the system. To visualize the solution to a system of inequalities, we graph the solution to each inequality in the system. The area where these solutions overlap is the region containing the solution to the system.

♦ Example 10.2: *(Washington, Section 17-5, Example 5)* Graph the region defined by the inequalities $y \geq -x - 2$ and $y + x^2 < 0$.

Solution:

1. First, solve each inequality for y.

$$y \geq -x - 2 \text{ and } y < -x^2$$

2. Next enter the corresponding equations in for function names and graph the two curves. Use the standard viewing rectangle. Let

$$Y_1 = -x - 2$$
$$Y_2 = -x^2$$

3. To shade the area between the two graphs, we first find the points of intersection of the two graphs (Procedure G15). These points are (−1, −1) and (2, −4).
4. Use Procedure G22 to shade in the area between the two graphs by letting the lower boundary be Y_1 and the upper boundary Y_2.

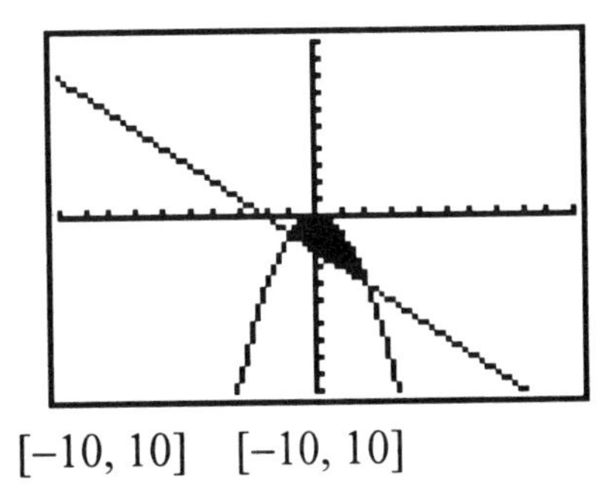

[−10, 10] [−10, 10]

Figure 10.2

The left most x-value is -1 and the right most x-value is 2. The graph appears in Figure 10.2.
The shaded area represents that set of points that satisfies both inequalities at the same time. When sketching this graph on paper it needs to be remembered that the curve $y = -x^2$ should be drawn as a dashed curve and the line $y = -x - 2$ as a solid line.

The corners of the region on a graph that represents the solution set of a system of inequalities are called the **vertices** of the region. The vertices may occur at the intersection of two graphs or at the x- or y-intercepts. The y-intercepts, in general, are easily found by letting $x = 0$ and solving for y. The x-intercepts of the function are the same as the roots of the corresponding equation. (The x-intercept of $y = 3x + 2$ is the same as the root of the equation $3x + 2 = 0$.) Procedures G11 and G15 describe the method of finding roots of equations and points of intersection of graphs.

Exercise 10.1

In the following exercises, use the graphing calculator to determine the solution set to the given inequality or system of inequalities. Based on your results, hand sketch a graph and shade in an area which represents the solution.
In Exercises 11-23, give all corner points (vertices) of the solution region and label these vertices on your hand sketched graph. Round all values to 3 significant digits.

1. $3x - 5y > 15$
2. $4x + 6y \le 12$
3. $4x + 7y - 18 \ge 0$
4. $-3x + 8y > 0$
5. $y + x^2 < 4$
6. $x^3 - 4x < y$
7. $y \le \sqrt{x+3}$
   $y \ge 0$
8. $y \le \dfrac{1}{x}$
   $y \ge 0, x \ge 1$
9. $y < 2 \log x$
   $y \ge 0$
10. $y < 5 \sin(2\pi x)$
    $x \ge 0, y \ge 0, x \le 1$
11. $15x + 13y \le 135$
    $x \ge 0, y \ge 0$
12. $23x + 19y < 243$
    $x \ge 0, y \ge 0$

13. $12x + 20y > 47$
    $x \geq 0,\ y \geq 0$

14. $0.25x - 1.47y \leq 23.45$
    $x \geq 0,\ y \geq 0$

15. $5x + 4y \leq 19$
    $y \leq 3,\ y \geq 0,\ x \geq 0$

16. $3x + 6y \geq 35$
    $y \leq x,\ y \geq 2$

17. $7x + 4y \leq 50$
    $x + 5y \leq 34$
    $y \geq 5,\ x \geq 0$

18. $x + y \leq 25$
    $3x + 5y \geq 44$
    $x \geq 0,\ y \geq 0$

19. $x + 5y \leq 65$
    $7x + 3y \geq 98$
    $x + 2y \geq 15$
    $x \geq 0,\ y \geq 0$

20. $2x + y \leq 14$
    $5x + 2y \geq 10$
    $-3x + 2y \geq 5$
    $x \geq 0,\ y \geq 0$

21. $-2x + 3y \leq 12$
    $y > x^2$

22. $y \geq 2^x$
    $y - 5x \leq 1$

23. $y < \sin x$
    $y \geq \cos x$
    $x \geq 0,\ x \leq 2\pi$

24. $y \leq \sqrt{25 - x^2}$
    $y - x^2 + 4 > 0$

In Exercises 25 - 28, set up the correct inequalities, use the graphing calculator as an aid to determine the region described, hand sketch a graph showing the region, and determine the vertices (corner points) of the region. Show the coordinates of the vertices on the hand sketched graph.

25. A cereal company manufactures x twelve ounce boxes of cereal and y eighteen ounce boxes every hour. The total number of boxes of cereal made each hour cannot exceed 185. Determine the region that represents the possible values for x and y.
26. A company sells two products, A and B. It makes a profit of \$5.50 for each unit of A and \$7.25 for each unit of B it sells. The total profit must be at least \$425. Determine the region that represents the possible number of units of A and B that need to be sold.
27. ZZZ Manufacturing Company has \$47,500 available for manufacturing machine parts. Part of this is to be spent on parts costing \$125 each to produce and part is to be spent on parts costing \$158 each to produce. At least 100 of the \$125 parts must be produced. Determine the region that represents the possible number of \$125 parts and \$158 parts the company is able to produce under these conditions.

28. A company produces two different types of computers. One type sells for $1250 and the other type sells $1570. It need to make at least $250,000 off of sales. It also has set a goal of selling at least 100 of the $1250 computers and 75 of the $1570 computers. Determine the region that represents the possible number of computers of each type that must be sold to meet these conditions..

◆ 29. *(Washington, Exercises 17-5, #33)* A telephone company is installing two types of fiber optic cables in an area. It is estimated that no more than 322 m of type A cable, and at least 175 m but no more than 475 m of type B cable, are needed. Graph the possible lengths of cable that are needed.

◆ 30. *(Washington, Exercises 17-5, #40)* A company makes computer parts A and B in each of two different plants. It costs $4000 per day to operate the first plant and $5000 per day to operate the second plant. Each day the first plant produces 125 of part A and 185 of part B, while the second plant produces 255 of A and 105 of B. Graph the region given the number of days for each plant to operate to meet these conditions. How many days should each plant operate to produce 2200 of each part and keep operating costs to a minimum?

Chapter 11

Polar Graphs

11.1 Graphing in Polar Coordinates
(Washington, Sections 21-9 and 21-10)

There are times when it is either easier or better to graph in coordinate systems other than in the familiar rectangular coordinate system. One important system is the polar coordinate system. In the polar coordinate system a point is located by using an angle, θ, measured from the positive x-axis and a distance, r, measured from the origin. The positive x-axis is known as the **polar axis** and the origin is known as the **pole**. Points in polar coordinates are expressed in terms of r and θ in the form (r, θ). (For conversion of coordinates see Procedure C12.)

Before using the graphing calculator to graph in polar coordinates, it is first necessary to set the calculator to polar graphing mode. After this is done a function may be entered and graphed.

<u>Procedure G23</u>. To graph in polar coordinates

1. Make sure the calculator is in correct mode for polar coordinates. (Procedure G17)
2. To enter polar equations:

   On the TI-82:
   Press the key Y=
   On the screen will appear:
   r1=
   r2=
   r3=
   (etc.)
   (up to 6 equations may be entered)

   On the TI-85:
   Press the key GRAPH.
   From the menu, select r(θ) =
   On the screen will appear: r1=
   (up to 99 polar equations may be entered as r1, r2, r3, ...)

   Functions may be turned on and off as with functions in rectangular coordinates (see Procedure G1, Note 3).
3. Enter the equation for r.
   For the variable θ

   On the TI-82:
   Press the key X,T,θ

   On the TI-85:
   Select θ from the menu.
4. Press or select GRAPH to graph the function or QUIT to exit.

Graphs in polar coordinates are plotted in order of increasing values of the angle θ rather than from left to right as in rectangular coordinates. Careful observation while the calculator is drawing the graph or using the trace function will indicate the order in which the points are plotted. As with regular graphs, we may use the zoom function of the calculator to zoom in on special points. Some adjustment of the viewing rectangle may be necessary when graphing in polar coordinates. In addition to the setting of minimum and maximum x- and y-values (Procedures G5 and G6), we now need to set values for θ.

Procedure G24. To change viewing rectangles values for θ.

With the calculator in the mode for polar coordinates:

1. Follow Procedure G5 to see the values for the viewing rectangle.
2. Enter new values for θmin, θmax, and θstep and press ENTER or move to the next item by using the cursor keys.
   The standard viewing rectangle values are

   θmin = 0, θmax = 6.28... (2π), and θstep = 0.13...$(\pi/24)$
   (or θmax = 360 and θstep = 7.5 if in degree mode).

   θstep determines how often points are calculated and may affect the appearance of the graph. Leaving θstep at approximately 0.1 is generally sufficient but θstep may have to be changed if there is a major change in the viewing rectangle. θmax should be sufficiently large to give a complete graph.
3. Press QUIT to exit.

The step size for θ will affect the appearance of the graph. If the step size is too large, a series of line segments may result rather than a smooth appearing curve. If the step size is too small, it may take too long to graph the curve. A step size of approximately 0.1 is usually sufficient. However, be aware that if we zoom in on a portion of the graph, this step size may be too large.

Example 11.1: Graph the polar function $r = 8 \cos \theta$, make a sketch of the graph, and give the polar coordinates, rounded to two decimal places, of the points where the graph crosses the x- and y-axes.

Solution:

1. Make sure the calculator is in the correct mode for polar graphing and use the standard viewing rectangle. Use $0 \le \theta \le 2\pi$.
2. Graph the function $r = 8 \cos \theta$ using Procedure G23.
   The graph should appear as an ellipse. If we change to a square viewing rectangle we see the graph is really a circle. The graph is in Figure 11.2.

3. With the graph on the screen, use the TRACE function to see points on the graph. As the cursor moves on the curve, the polar coordinates are given at the bottom of the screen. The points where the graph intersects the axes (in polar coordinates) are (8.00, 0) and $(0, \frac{\pi}{2})$. Care must be taken in obtaining values near the pole, since values may vary greatly with small movements of the cursor. Values obtained for intersection points should be tried in the equation to be sure they are correct.

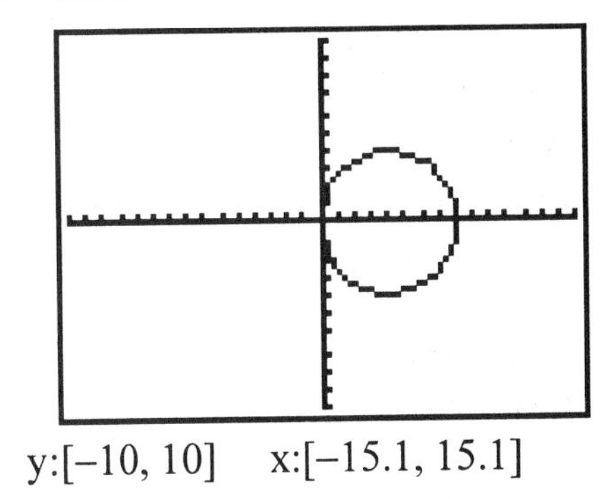

y:[−10, 10] x:[−15.1, 15.1]

Figure 11.2

Certain forms of functions in polar coordinates, as in rectangular coordinate, give particular types of graphs. We now consider the standard forms of some graphs.

Roses

A rose has the standard forms

$$r = a \sin (b\theta) \quad \text{or} \quad r = a \cos(b\theta)$$

In the exercises, we will see the effect of changing the values of a and b. The value a has an effect on the size of the graph and the value b determines the number of leaves or petals on the rose. A special cases of these equations may give a rose of one leaf (a circle).

Limaçons

A limaçon is formed by equations of the form

$$r = a + b \sin t \quad \text{or} \quad r = a + b \cos t$$

Certain special cases of limaçons result in curves called cardioids. The relative sizes of a and b determine different limaçons.

Conics

Various parabolas, ellipses, and hyperbolas are graphed by equations of the form

$$r = \frac{a}{b + c\sin\theta} \quad \text{or} \quad r = \frac{a}{b + c\cos\theta}$$

The type of curve and the size will be determined by the relative values of a, b, and c. For such a conic, one focus will always be at the pole.

Spirals

Certain spirals are graphed by functions of the form

$$r = a\theta^b \quad \text{or} \quad r = a^{b\theta}$$

The values of a and b determine the relative size and shape of the spiral.

Exercise 11.1

Use the calculator to graph the following in polar coordinates. Points should all be expressed in polar coordinates. For best results, use a square viewing rectangle and $0 \le \theta \le 2\pi$.

In problems 1-10, (a) determine what type of graph is obtained and draw a sketch of the graph on your paper, (b) give the maximum distance of a point on the graph from the pole, and (c) give the points, in polar coordinates, (to one decimal place) where the graph crosses the x- and y-axis.

1. $r = 4 \cos \theta$
2. $r = -4 \sin \theta$
3. $r = -6 \cos 2\theta$
4. $r = 8 \sin 2\theta$
5. $r = 6 + 4 \sin \theta$
6. $r = 5 + 5 \cos \theta$
7. $r = \dfrac{1}{5 + \sin\theta}$
8. $r = \dfrac{1}{4 + 4\cos\theta}$

9. $r = 4\theta^2$
10. $r = 1.5(2^\theta)$

Do problem 11-20 as indicated.

11. Graph $r = \cos\theta$, $r = 2\cos\theta$, and $r = 4\cos\theta$ on the same set of axes. What kind of graphs are obtained? What happens to a graph of the form $r = A\cos\theta$ as A changes?
12. Graph $r = 5\sin 2\theta$ and $r = 5\cos 2\theta$ on the same axes. Then graph $r = 5\sin 4\theta$ and $r = 5\cos 4\theta$ on the same axes. What is the difference between using $\sin\theta$ and using $\cos\theta$?
13. Graph $r = 8\sin 2\theta$, $r = 8\sin 4\theta$, and $r = 8\sin 6\theta$. Describe the graphs obtained. How does the graph of $r = 8\sin B\theta$ change as B changes? Does this hold for the equation $r = 8\sin 3\theta$?
14. Graph $r = 6\sin 3\theta$, $r = 6\sin 5\theta$, and $r = 6\sin 7\theta$. Describe the graphs obtained. How do these graphs differ from those in problem 13? What is the relation of the number of leaves and the number B in the equation $r = A\cos B\theta$?
15. Graph $r = 5 - 3\sin t$. What type of graph is obtained? (Zoom in on the point where $\theta = \frac{\pi}{2}$.) Graph $r = 4 - 4\sin\theta$, and $r = 3 - 5\sin\theta$. (Use the square viewing rectangle.) How do these graphs differ? How do the relative sizes of A and B effect the graphs of the equation $r = A - B\sin\theta$?
16. Graph $r = 6 - 2\cos\theta$, $r = 6 - 6\cos\theta$, and $r = 2 - 6\cos\theta$. How do these graphs differ? How do the relative sizes of A and B effect the graphs of the equation $r = A - B\cos\theta$?
17. Graph $r = \dfrac{10}{5-3\sin\theta}$, $r = \dfrac{10}{3-3\sin\theta}$, and $r = \dfrac{10}{3-5\sin\theta}$. What kind of graphs are obtained? How do these graphs differ? How do the relative sizes of B and C effect the graphs of the equation $r = \dfrac{A}{B-C\sin\theta}$?
18. Graph $r = \dfrac{10}{3-2\cos\theta}$, $r = \dfrac{10}{2-2\cos\theta}$, and $r = \dfrac{10}{2-3\cos\theta}$. What kind of graphs are obtained? How do these graphs differ? How do the relative sizes of B and C effect the graphs of the equation $r = \dfrac{A}{B-C\cos\theta}$?
19. Graph the two functions $r = 5\sin\theta$ and $r = 5\cos\theta$ on the same set of axes. Estimate the points of intersection to two decimal places.
20. Graph the two functions $r = -3\cos\theta$ and $r = 3 + 5\sin\theta$ on the same set of axes. Estimate the points of intersection to two decimal places.

21. Graph $r = 0.2\theta^2$. What type of graph is obtained? Give the first 3 points (for $\theta > 0$) that the graph crosses the x- and y-axes.
22. Graph $r = 2^{3\theta}$. What type of graph is obtained? Give the first 3 points (for $\theta > 0$) that the graph crosses the x- and y-axes.
23. The shape of a cam in a piece of machinery is given by the equation $r = 8.7 - 5.2 \cos \theta$. Graph this function and sketch the graph on your paper. Determine the minimum and maximum distance of a point on the edge of the cam from the pole.

◆ 24. *(Washington, Exercises 21-9, # 40)* The polar equation of the path of a weather satellite about the earth is

$$r = \frac{4824}{1 + 0.145\cos\theta}$$

where r is the distance from the center of the earth and is measured in miles. The path is an ellipse with the center of the earth at one foci. Graph on the calculator and sketch the graph on your paper. Assume the earth is a circle whose radius is 3960 miles. What is the maximum distance the satellite is above the earth's surface and what is the minimum distance?

◆ 25. *(Washington, Exercises 21-10, #33)* An architect designs a patio which is shaped such that it is described as the area within the polar curve $r = 3.25 - 3.25\sin\theta$. Measurements are in meters. Graph on your calculator and sketch the shape of the patio on your paper. What is the width of the patio along the x-axis? Find the angle between the line from the pole to the point where $r = 3.25$ and the line from the pole to the point where $r = 6.00$.

◆ 26. *(Washington, Exercises 21-10, #34)* A missile is fired at an airplane and is always directed toward the airplane. The missile is traveling at twice the speed of the airplane. An equation that describes the distance r between the missile and the airplane is

$$r = \frac{72\sin\theta}{(1 - \cos\theta)^2}$$

where θ is the angle between their directions. The plane is assumed to be at the pole at all times. Graph this for $\frac{\pi}{4} \le \theta \le \pi$. Show the graph on your paper. What is the distance of the missile from the plane when $\theta = \frac{\pi}{2}$? When $\theta = \pi$?

Chapter 12

Statistics

12.1 Statistical Graphs and Basic Calculations
(Washington, Sections 22.1, 22.1, and 22.3)

Statistics is the collection, organization, and interpretation of numerical data. This section discusses how data is organized, takes a first look at the interpretation of data, and shows how the graphing calculator may be used to help analyze data.

If the amount of data is relatively small, it is not difficult to look at the data as individual items. However, if the amount of data is large, with many values, each occurring a large number of times, then the data needs to be grouped into what is called a frequency distribution. A **frequency distribution** is a table giving each individual data value and the frequency of each data value. The **frequency** is the number of times a particular data value occurs in that set of data.

To visualize data given in a frequency distribution, it is helpful to see a graph of the data. Three different types of graphs are considered: a histogram, a scatter graph, and a frequency polygon. A **histogram** is a graph in which the each data item or group of data items is represented by a rectangle. All the rectangles are the same width and the height of each rectangle represents the frequency of that particular data value. This type of graph is commonly known as a bar graph. In a **scatter graph** points are plotted representing the frequency of each of the data values. A **frequency polygon** (or **xyLine graph**) is a graph in which the points of a scatter graph are connected by straight line segments.

Before working with statistical data on the calculator, we first need to organize the data and store it in the calculator.

<u>Procedure S1</u>. To enter or change single variable statistical data.

On the TI-82 and TI-85 calculators data is stored as lists of numbers. (See also Procedure C7)

1. Press the key STAT
2. Enter edit mode.

   On the TI-82:
   With EDIT highlighted in the top row, select 1:Edit...

   On the TI-85:
   Select EDIT from the menu.
3. Enter the data values and frequencies.

All frequencies must be entered as integers.

On the TI-82:
Enter the data values under one list name. After entering every data value, press ENTER. Enter the related frequencies under a second list name. After entering each frequency, press ENTER. Make sure the frequencies are entered in the same order as the data items so that they correspond to the correct item.

On the TI-85:
The calculator will ask for an xlist name and a ylist name. The xlist is for the data values and the ylist is for the frequencies. Enter a name (up to eight characters) for xlist and press ENTER and enter a name for ylist and press ENTER. Enter each data value as an x-value and the corresponding frequency as a y-value. (y_1 is the frequency for data item x_1, y_2 for x_2, etc.)

Note: When there are intermediate data values with a frequency of zero, it is best, for graphing purposes, to enter these data values and the zero frequencies.

4. To edit data.

   On the TI-82:
   Move the cursor to the values to be changed, make the change, and press ENTER. As the data item in a list is highlighted, the value of that item appears at the bottom of the screen.

   On the TI-85:
   Move the cursor to the values to be changed and make the desired change.

5. To exit, after all data have been entered and the cursor appears on the next value, press QUIT.
6. To view a list, see Procedure C8.

After the data is stored, graphs may be drawn and calculations performed. When working with the calculator, the data values are the x-values and are plotted on the horizontal axis and the frequencies of the data values are the y-values and are plotted on the vertical axis.

Procedure S2. To graph single variable statistical data.

1. Make sure the viewing rectangle is appropriate for the data to be graphed, that the graph display has been cleared, and that all functions in the function table have been turned off.
2. Be sure the data to be graphed is stored in lists in the calculator. (Procedure S1)
3. Define the plot and draw the graph:

   On the TI-82:
   (a) Press the key STAT PLOT
   (b) Select one of the three plots.

   On the TI-85:
   Press the key STAT
   From the menu, select DRAW

The screen will show the plot number selected, followed by several options. The highlighted options are active. To highlight an option, move the cursor to that option and press ENTER.

(c) Highlight ON.

(d) Highlight the type of graph. The first symbol following the word Type is for a scatter plot, the second for a xyLine, the third for a box plot, and the fourth for a histogram.

(e) For Xlist, highlight the name of the list of data values.

(f) For Ylist, highlight the name of the list of frequencies.

(g) For Mark, highlight the symbol to be used in plotting data.

(h) To exit, press QUIT

(i) To see the graph, press GRAPH.

Turn the STAT PLOT off when finished.

From the secondary menu, to draw a graph, select:
HIST for a histogram.
SCAT for a scatter graph
xyLine for a frequency polygon.

Note: For a histogram the value of Xscl will determine width of bars.
One type of graph should be cleared before displaying another.

Procedure S3. To clear the graphics screen.

1. All functions should be deleted or turned off.
2. Clear the screen:

On the TI-82:
Press the key DRAW.
With DRAW highlighted in the first row, select 1:ClrDraw.
Press ENTER.

On the TI-85:
Press the key STAT
From the menu, select DRAW.
From the secondary menu, select CLDRW.

Example 12.1: Given the following set of numbers, construct a frequency distribution, enter this distribution into the calculator, and graph the data as a scatter graph, as a histogram, and as a frequency polygon.

5, 7, 3, 6, 5, 9, 5, 6, 7, 3, 4, 8, 6, 7, 5, 8, 3, 4, 6, 6

Solution:

1. Notice that the smallest data value is 3 and the largest value is 9. To construct a frequency distribution, make a table of two rows. In the first row, list each of the integers from 3 through 9. In the second row, below the corresponding value in the first row, give the frequency (number of times) that each value occurs. In this case the frequency distribution is

x(data):	3	4	5	6	7	8	9
y(freq.):	3	2	4	5	3	2	1

2. Enter this data into the calculator (see Procedure S1).
3. Before drawing the graphs, make sure the viewing rectangle is appropriate and that previous graphs are cleared from the screen (see Procedure S3). Since the x-values vary from 3 to 9, make Xmin = 0, Xmax = 11 and Xscl = 1 and since the y-values go from 1 to 5, make Ymin = 0, Ymax = 8, and Yscl = 1. These values are taken to allow a little extra space around the edges of the graph.
4. Use Procedure S2 to graph the data. Clear each graph before doing another. The scatter graph is in Figure 12.1, the histogram in Figure 12.2 and the frequency polygon in Figure 12.3.

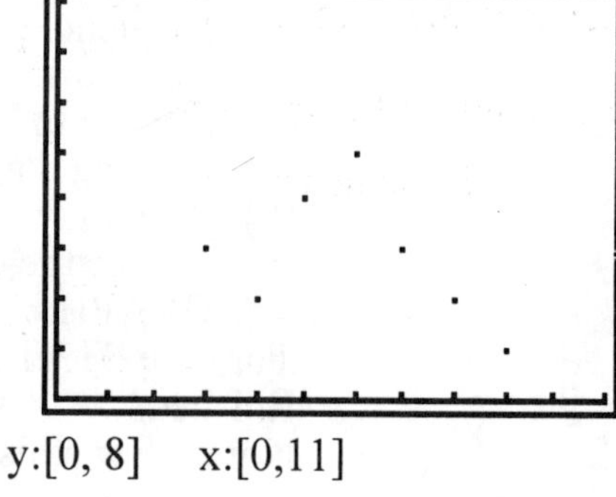

Figure 12.1

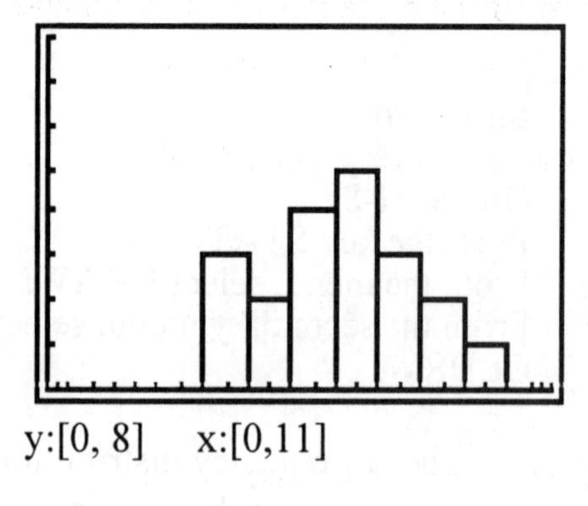

Figure 12.2

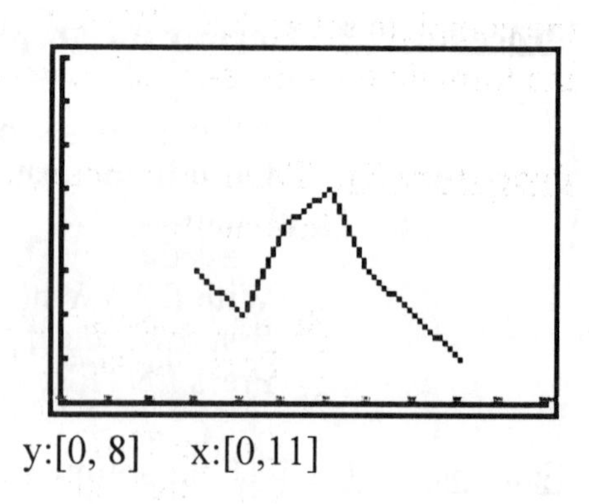

Figure 12.3

The graphs of Example 12.1 give different representations of the same data. The choice of which to use will depend on personal preference and the given situation.

We next proceed with additional analysis of data. Two quantities that are of particular importance when working with data are the mean and the standard deviation of the data. The mean is the sum of all the x-values (data values) divided by the total number of x-values. This is represented symbolically as

$$\bar{x} = \frac{\sum x_i}{n} \quad \text{or} \quad \bar{x} = \frac{\sum f_i x_i}{\sum f_i}$$

where $\sum x_i$ means the sum of all the x-values (individual values are represented by x_i, for i = 1, 2, 3,...), n is the total number of x-values, and f_i represents the frequency of each x-value x_i. The sum of all the x-values is equal to the sum of the products of each x-value with its frequency, $\sum f_i x_i$. The total number, n, of all x-values is equal to the sum of the frequencies, $\sum f_i$.

The standard deviation is calculated from one of the following formulas:

$$\sigma = \sqrt{\frac{\sum (x_i - \bar{x})^2}{n}} \quad \text{or} \quad s = \sqrt{\frac{\sum (x_i - \bar{x})^2}{(n-1)}}$$

The formula for σ (the Greek letter sigma) is used to calculate standard deviation when every element of the data under study is used, while the formula for s is used to determine the standard deviation when a sample of data is used as an approximation to the complete set of data. We will call the formula for σ the σ-standard deviation and the formula for s the s-standard deviation. The two values will differ slightly.

Procedure S4. To obtain mean, standard deviation and other statistical information.

Be sure one-variable data has been stored as in Procedure S1.

1. Press the key STAT
2. Obtain the statistical information:

On the TI-82:
(a) Highlight CALC in the top row.

On the TI-85:
(a) From the menu, select CALC.

(b) From the menu, select 3:SetUp
(c) Under 1-Var Stats, select the list name to be used for the data values and the list name to be used for the frequencies. (If each has a frequency of one, select 1.)
(d) Press QUIT
(e) Again press STAT and highlight CALC in the top row.
(f) From the menu, select 1:1-Var Stats, then press ENTER

(b) Make sure the xlist name is correct and press ENTER and the ylist name is correct and press ENTER.
(c) From the secondary menu, select 1-VAR.

A list of statistical quantities appear.

$\overline{x}$ is the mean of all the x-values
$\sum x$ is the sum of all the x-values
$\sum x^2$ is the sum of all the squares of the x-values
Sx is the s-standard deviation of the x-values
σx is the σ-standard deviation of the x-values
n is the total number of x-values

3. Clear the screen.
   On the TI-82:
   Press CLEAR to clear the screen.

   On the TI-85:
   Press QUIT to exit.

<u>Example 12.2</u>: Find the mean and standard deviation of the data in Example 12.1.

<u>Solution</u>:

1. Check to make sure the data is still stored in the calculator. If it is not, it will have to be reentered. (Procedure S1)
2. Use Procedure S4 to obtain the statistical data. This screen tells us that the mean of the x-values is 5.65, the s-standard deviation is 1.73 and that the σ-standard deviation is 1.68 (values rounded to two decimal places) and that we had a total of 20 data values.

Before entering new data it is always a good idea to clear old statistical data from the memory of the calculator. Otherwise, the two sets of data may become intermixed and the results may be in error.

<u>Procedure S5</u>. To clear statistical data from memory.

1. Press the key: MEM.
2. From the menu, select Delete.
3. Then select, List.

4. Use the cursor keys to move the indicator to the name of the list to be deleted and press ENTER.
5. Press QUIT to exit.

Note: On the TI-82, data may also be deleted by: Press the key STAT, select 4:ClrList, enter the names, separated by commas, of the lists to be deleted (by pressing the keys for those names), then press ENTER.

Two other important quantities when considering a set of data are the median and the mode of the data. The **median** is that value which lies in the middle when the data is arranged in numerical order. If there are an odd number of values, the median is the middle value. If there are an even number of values, the median is the average of the two middle values. The **mode** is that value that occurs most often. The median and mode are generally not displayed by graphing calculators and must be found by hand.

The quantities mean, median, and mode are called measures of central tendency. They are an indication as to where the middle of the data is located. The standard deviation is an indication as to the extent the data deviates from the mean on the average. These quantities are used extensively for comparison and analysis of data.

Example 12.3: Find the median and mode for each of the following sets of data.

(a) 3, 4, 9, 3, 3, 5, 2, 8, 4
(b) 12, 16, 10, 14, 14, 15, 18, 15, 14, 16
(c) The data in Example 12.1 .

Solution:

1. First, for parts (a) and (b) rearrange the values in numerical order. For part(c), we will use the frequency distribution of Example 12.1.
   (a) 2, 3, 3, 3, 4, 4, 5, 8, 9
   (b) 10, 12, 14, 14, 14, 15, 15, 16, 16, 18
   (c) (value) x: 3 4 5 6 7 8 9
       (freq.) y: 3 2 4 5 3 2 1
2. (a) For the first set, there is an odd number of values. Therefore the median is the middle value or 4. The values that occurs most often is 3, thus, the mode is 3.
   (b) For the second set, there is an even number of values. Therefore the median is the average of the two middle values (14 and 15) or 14.5. The value which occurs most often is 14 and 14 is the mode.
   (c) In the case of the third set, the total number of values (the sum of the frequencies) is 20. Thus, the median will be the average of the 10th and 11th values. To obtain these, add the

frequencies starting from one end. Both values are 6. Thus, the median is 6. The value which occurs most often in this case is also 6, which is the mode.

Exercise 12.1

In Exercises 1-5, 7 and 8, (a) construct a frequency distribution of the data and enter the data into the calculator, then obtain (b) a histogram, (c) a scatter graph, (d) a frequency polygon, (e) the mean, (f) the median, (g) the mode, and (h) the s-standard deviation. Sketch graphs for parts (b), (c), and (d) on your paper.

1. For a set of quiz scores (10 max. score):
   6, 7, 10, 8, 9, 8, 7, 7, 5, 10, 10, 8
2. For a set of quiz scores (10 max. score):
   9, 10, 8, 8, 7, 5, 6, 9, 9, 10, 8
3. For a set of test scores (100 max. score):
   85, 90, 100, 75, 85, 95, 90, 80, 80, 90, 100
4. For a set of test scores (100 max. score):
   65, 75, 85, 100, 85, 75, 85, 95, 90, 90, 95, 85
5. For a set of test scores (100 max. score):
   75, 85, 80, 90, 90, 85, 75, 100, 100, 90, 95, 75
6. Based on your results for problems 3 and 5, which class did the best on the test? On what do you base you answer?
7. ♦ *(Washington, Exercises 22-1, #17)* In testing a computer system, the number of instructions it could perform in 1 ns was measured at different points in a program. The number of instructions were as follows:

   19, 21, 24, 25, 22, 21, 23, 24, 18, 18, 19, 22, 24, 22, 19

8. ♦ *(Washington, Exercises 22-1, #27)* The life of a certain type of battery was measured for a sample of batteries with the following results (in number of hours):

   34.2, 30.5, 32.3, 35.6, 31.0, 28.7, 29.4, 30.2, 32.7, 25.6, 30.9, 28.7, 36.3

In Exercises 9-12, find (a) the mean, (b) the s-standard deviation, and (c) the number of data values that lie between the mean minus one standard deviation and the mean plus one standard deviation ($\overline{x} \pm s$).

9.

value	frequency	value	frequency
21.2	2	23.2	6
21.6	4	23.5	5
21.8	5	23.8	5
21.9	8	24.0	3
22.0	7	24.2	2
22.3	8	24.5	1
22.8	11		

10.

value	frequency	value	frequency
50	2	62	8
51	5	64	7
52	8	65	6
53	9	67	3
55	12	69	2
57	12	70	2
59	10	71	1
60	9	72	1

♦ 11. *(Washington, Section 22-3, #25 and #26)* The weekly salaries (in dollars) for the workers in a small factory are as follows:

355, 445, 385, 325, 285, 410, 305, 485, 375, 510, 410, 335, 350, 400

♦ 12. *(Washington, Section 22-3, #24)* In testing a braking system, the distance required to stop a car from 70 mi/h was measured in several trials. The results are in the following table:

Stopping distance	155-159	160-164	165-169	170-174	175-179	180-184
Times car stopped	4	17	35	37	23	8

13. A worker measures the diameters of a sample of gaskets and obtains the following frequency distribution table giving the number of times each diameter (in inches) was observed. Find the mean, median, mode, standard deviation and the number of data values that lie between the mean minus one standard deviation and the mean plus one standard deviation ($\overline{x} \pm s$).

diam.	frequency	diam.	frequency
3.25"	1	3.31"	23
3.26"	4	3.32"	25

3.27"	8	3.33"	22
3.28"	10	3.35"	16
3.29"	14	3.36"	13
3.30"	20	3.37"	3

14. A lab technician measures the weight (in ounces) of cereal in one pound boxes and obtains the following frequency distribution table giving the number of times each weight was observed. Find the mean, median, mode, standard deviation and the number of data values that lie between the mean minus one standard deviation and the mean plus one standard deviation ($\overline{x} \pm s$).

Wt.	frequency	Wt.	frequency
15.5	2	16.1	25
15.6	4	16.2	23
15.7	6	16.3	12
15.8	10	16.4	8
15.9	21	16.5	5
16.0	31	16.6	1

15. A consumer testing company has a technician test a sample of light bulbs for length of life. She compiles the data into the following frequency distribution giving the number of bulbs in the sample which lasted the given number of hours. Find the mean, median, mode, standard deviation and the number of data values that lie between the mean minus one standard deviation and the mean plus one standard deviation ($\overline{x} \pm s$).

time	frequency	time	frequency
1150	3	1200	125
1155	10	1205	148
1160	35	1210	118
1165	28	1215	85
1170	57	1220	112
1175	65	1225	87
1180	72	1230	67
1185	88	1235	45
1190	92	1240	10
1195	102	1245	5

16. A worker measures the diameter of a samples of bolts taken from a bin containing 10,000 bolts. He measures the diameter (in millimeters) of each bolt in the sample and arrives at the following frequency distribution. Find the mean,

median, mode, standard deviation and the number of data values that lie between the mean minus one standard deviation and the mean plus one standard deviation ($\bar{x} \pm s$).

diam.	frequency	diam.	frequency
10.95	1	11.06	24
10.96	4	11.07	15
10.97	2	11.08	16
10.98	12	11.09	11
10.99	24	11.10	8
11.00	36	11.11	5
11.01	43	11.12	3
11.02	48	11.13	0
11.03	52	11.14	2
11.04	49	11.15	1
11.05	44		

♦ 17. *(Washington, Exercises 22-1, # 31)* Toss three coins 100 times and tabulate the number of heads that appear for each toss. Construct a frequency distribution table for your data. Enter the data on your calculator and construct a histogram and a frequency polygon. Determine the mean and standard deviation of your data. What percentage of your values occur between the mean minus one standard deviation and the mean plus one standard deviation ($\bar{x} \pm s$)? Compare your data with others in the class. Discuss the differences obtained.

18. Simulate the tossing of two dice and finding their sum by entering and running the following program on your calculator.

```
PROGRAM: DICETOSS
int (6 rand)+1 → X
int (6 rand)+1 → Y
X + Y → S
Disp S
```

[rand is a rand number generator and is found on the calculator by pressing the MATH key and selecting PRB. To get →, press the key STO▷.]
Run this program 50 times by repeatedly pressing ENTER before pressing any other key. Keep track of the data and form a frequency distribution. Enter the data on your calculator and construct a histogram and a frequency polygon. Determine the mean and standard deviation of your data. What percentage of your values occur between the mean minus one standard deviation and the mean plus one standard deviation ($\bar{x} \pm s$)? Compare your data with others in the class. Discuss the differences obtained.

12.2 Regression - Finding the Best Equation
(Washington, Sections 22-4 and 22-5)

Two variables are frequently related to each other in such a way, that as one changes, this causes a change in the other variable. This relationship can be expressed by writing one variable as a function of a second variable. The first variable is considered the **dependent variable** and the second the **independent variable**. For example, from Ohm's law, we know that in a simple direct current circuit containing a 12 ohm resistor the voltage equals 12 times the current in the circuit. If we let E represent the voltage and I represent the current then we arrive at the equation $E = 12I$. Here the E is considered the dependent variable and is expressed as a function of I which is considered the independent variable.

Many times the equation which relates two variables is not known and must be found from a collection of known data or by collecting data experimentally. A first step in attempting to determine such an equation is to graph the data. This will often indicate the type of equation, such as linear, logarithmic, etc., that will best represent the data. We will consider four types of equations that are used to represent data: linear, logarithmic, exponential, and power equations. The type of equation that best represents a given set of data is determined by trying different graphs and seeing when a straight line is obtained. These graphs often involve logarithms of one or both variables and may be graphed by plotting the logarithm of the variable or by using semi-logarithm or logarithmic paper. The type of equation for each type of graph is:

Type of Equation	Data gives a straight line when
Linear	y graphed as function of x
Logarithmic	y graphed as function of log x
Exponential	log y graphed as function of x
Power	log y graphed as function of log x

From the graph we could, by using a few well selected points, come up with an equation which "fits" the graph. However, we can find a better equation, by making use of a mathematical method called **least squares regression.**

Many graphing calculators have several mathematical methods built in which may be used to determine an equation that best "fits" a given set of data. Although many calculators are capable of determining other types of equations, we will only consider the four types of regression equations listed here. After the data has been properly entered into the calculator, the calculator gives the constants in the corresponding regression equations.

Type of Regression	Formula
Linear Regression	$y = a + bx$
Logarithmic Regression	$y = a + b \ln x$
Exponential Regression	$y = ab^x$
Power Regression	$y = ax^b$

In these formulas, the constants (a and b) are calculated for a given set of data, x represents the independent variable and y the dependent variable. For each Regression calculation, the calculator will also provide a value for the correlation coefficient, r, which is an indication as to how well a given equation "fits" a set of data. The closer r is to 1 or to −1 the better the equation represents the given set of data.

Data of this type is two variable statistical data and is entered into the calculator in much the same way as one variable statistical data.

Procedure S6. To enter or change two variable statistical data.

On the TI-82 and TI-85 calculators, data is stored as lists of numbers. (See Procedure C7)

1. First press the STAT.
2. Enter edit mode.
   On the TI-82:
   With EDIT highlighted in the top row, select 1:Edit...
   On the TI-85:
   Select EDIT from the menu.
3. Enter the x-values as one list and the y-values as a second list:
   On the TI-82:
   Enter the x-values under a list name. After every value is entered, press ENTER.
   Enter the related y-values under a second list name. After each value is entered, press ENTER.
   Make sure the y-values are entered in the same order as the x-values so that they correspond.
   On the TI-85:
   The calculator will ask for an xlist name and a ylist name. The xlist is for the x-values and the ylist is for the y-values. Enter a name (up to eight characters) for xlist and press ENTER and enter a name for ylist and press ENTER .
   Enter the first x-value and the corresponding y-value, the second x-value and the second y-value, etc.
4. To edit data.
   On the TI-82:
   Move the cursor to the values to be changed, make the change, and press ENTER. As the data item in a list is highlighted, the value of that item appears at the bottom of the screen.
   On the TI-85:
   Move the cursor to the values to be changed and make the desired change.

5. To exit, after all data have been entered and the cursor appears on the next value, press QUIT.
6. To view a list, see Procedure C8.

After the data is stored in the calculator, we are ready to determine regression equation.

Procedure S7. To determine a regression equation.

1. Be sure two-variable data has been stored as in Procedure S6.
2. Obtain the proper menu:

   On the TI-82:
   (a) Press the key STAT
   (b) Highlight CALC in the top row.
   (c) From the menu, select 3:SetUp.
   (d) Under 2-Var Stats highlight the list name to be used for the x-values and the list name to be used for the y-values. For Freq highlight 1 unless the number of times various points occur are listed under a third list name.
   (e) Press QUIT.
   (f) Again press the key STAT and highlight CALC in the top row.

   On the TI-85:
   (a) Press the key STAT
   (b) From the menu, select CALC.
   (c) Make sure the xlist name is correct and press ENTER and the ylist name is correct and press ENTER.

3. Select the type of regression equation. (The form of the menu item for the TI-82 is on the left and for the TI-85 is on the right.)
   (a) For linear regression ($y = a + bx$), select:
   9:LinReg(a+bx) LINR
   (b) For logarithmic regression ($y = a + b \ln x$), select
   0:LnReg LNR
   (c) For Exponential Regression ($y = ab^x$), select
   A:ExpReg EXPR
   (d) For Power Regression ($y = ax^b$), select
   B:PwrReg PWRR
   (e) For Quadratic Regression ($y = ax^2 + bx + c$), select
   6:QuadReg P2REG
   (f) For Cubic Regression ($y = ax^3 + bx^2 + cx + d$), select
   7:CubicReg P3REG
   (g) For Quartic Regression ($y = ax^4 + bx^3 + dx^2 + ex + f$), select
   8:QuartReg P4REG

   On the TI-82, after making the selection, press ENTER.

The calculator gives values for constants in the equations and the correlation coefficient.

◆ Example 12.4: *(Washington, Section 22-4, Example 5)* In a research project to determine the amount of a drug that remains in the bloodstream after a given dosage, the amounts y (in mg of drug/dL of blood) were recorded as in the following table. Find the linear regression equation which represents the following set of data.

t (hours)	1.0	2.0	4.0	8.0	10.0	12.0
y (mg/dL)	7.6	7.2	6.1	3.8	2.9	2.0

Solution:

1. First use Procedure S5 to clear any previously entered data from the statistics memory.
2. Then, use Procedure S6 to enter this two-variable data into the calculator.
3. Now, use Procedure S7 to find the linear regression equation.
4. The screen shows the values a = 8.15.., b = –0.523.., and r = –0.999. This means that the equation of the line (of the form y = a + bx) which best represents this set of data is y = 8.15 – 0.523x. The r value of –0.990 is very close to –1 indicating a good fit to the data.

The data in Example 12.4 is to two significant digit accuracy. When writing the coefficients for the equation, we use one additional significant digit to insure an accuracy of two significant digits in calculated values of x or y.

Two variable statistical data may be graphed on the calculator screen by making a scatter graph. This allows us to compare visually the data and the corresponding regression equation.

Procedure S8. To graph data and the related regression equation.

1. Make sure the viewing rectangle is set properly to graph the set of data and that the graph display has been cleared, and that all functions in the function table have been removed or turned off.
2. If the data has not already been entered into the calculator, it will need to be entered (Procedure S6) and the linear regression equation calculated (Procedure S7).

3. Store the regression equation for a function name and graph by:

On the TI-82:

(a) Press the key Y=
(b) Move the cursor after an unused function name.
(c) Press the key VARS
(d) From the menu, select 5:Statistics...
(e) Move the highlight in the top row to EQ
(f) Select 7:RegEQ
(g) Press the key GRAPH

On the TI-85:

(a) Press the key GRAPH
(b) From the menu, select y(x)=
(c) Move the cursor after an unused function name.
(d) Press the key VARS
(e) From the menu, select STAT (after pressing MORE twice).
(f) Move the mark on the left to RegEq
(g) Press ENTER
(h) Select Graph from the graph menu.

4. Define the plot and draw a scatter graph:

On the TI-82:

(a) Press the key STAT PLOT
(b) Select one of the three plots.

The screen will show the plot number selected, followed by several options. The highlighted options are active. (To make an option active, move the cursor to that option and press ENTER.)

(c) Highlight the option ON.
(d) Highlight the type of graph. The first symbol following the word Type is for a scatter plot.
(e) For Xlist, highlight the name of the x-values.
(f) For Ylist, make the name of the y-values active.
(g) For Mark, make active the symbol to be used in plotting data.
(h) To exit, press QUIT
(i) To see the graph, press GRAPH.

On the TI-85:

(a) Press the key STAT
(b) From the menu, select DRAW
(c) From the secondary menu, to draw a graph, select: SCAT for a scatter graph.

Example 12.5: Graph the data and the linear regression equation for the data of Example 12.4.

Solution:

1. If the data from Example 12.4 is still in your calculator continue with the next step, otherwise, reenter the data and calculate the regression equation (see Procedure S7).
2. Clear the graphics screen (Procedure S3) and make sure the viewing rectangle is set correctly. For this data, set the following values: Xmin = 0, Xmax = 14, Xscl = 1, Ymin = 0, Ymax = 10, Yscl = 1. This allows for some space around the graph and shows the graph relative to the x- and y-axis.
3. Use Procedure S8 to graph the data and the related linear regression equation. The graph i in Figure 12.4.

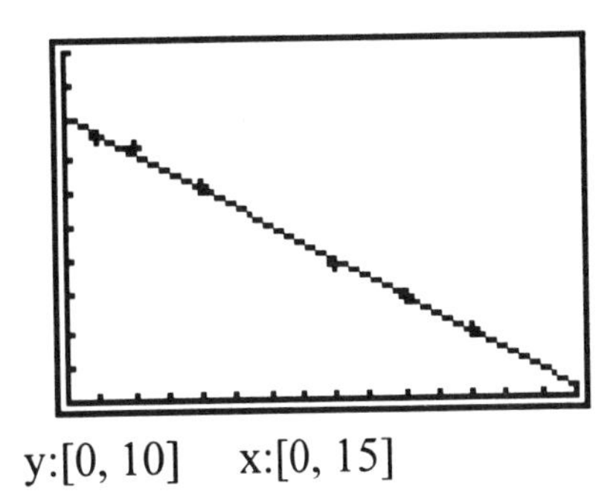

y:[0, 10] x:[0, 15]

Figure 12.4

Other regression formulas may be used in a similar manner. If it is not known which formula would give the best fit, it may be helpful to do a calculation for each formula and note the corresponding values for the coefficient of regression.

Exercises 12.2

In Exercises 1-10, find the regression equation, sketch a graph of the resulting equation on your paper and show the data points on the graph.

1. Find the linear regression equation which best "fits" the data. Also, give the correlation coefficient. Plot the equation and the data on the same screen. How well does the equation "fit" the data?

x	25	50	59	80	40	41	49	52
y	11	14	18	24	12	13	15	16

2. Find the linear regression equation which best "fits" the data. Also, give the correlation coefficient. Plot the equation and the data on the same screen. How well does the equation "fit" the data?

x	0	2	45	64	79	93	95	127	141
y	0	0.3	21.5	29.8	36.9	43.4	43.9	62.1	68.4

◆ 3. *(Washington, Exercises 22-4, #9)* The pressure p was measured along an oil pipeline at different distances from a reference point, with results as shown in the following table. Find the linear regression equation which best "fits" the data. Also, give the correlation coefficient. Plot the equation and the data on the same screen.

x (ft)	0	112	220	320	405
p (lb/in^2)	651	628	603	585	568

◆ 4. *(Washington, Exercises 22-4, #10)* The heat loss L per hour through various thicknesses of a particular type of insulation was measured as shown in the following table. Find the linear regression equation which best "fits" the data. Also, give the correlation coefficient. Plot the equation and the data on the same screen. How well does the equation "fit" the data?

x (in)	2.8	4.1	4.9	6.2	7.4	7.9
y (Btu)	5910	4750	3880	3130	2440	1670

5. Find the logarithmic regression equation which best "fits" the data. Also give the correlation coefficient. Plot the equation and the data on the same screen. How well does the equation "fit" the data? Find y, when x = 10.

x	2.4	3.5	7.6	8.4	9.9	11.7	14.6
y	4.90	5.40	7.10	7.35	7.48	7.80	8.48

6. Find the logarithmic regression equation which best "fits" the data. Also give the correlation coefficient. Plot the equation and the data on the same screen. How well does the equation "fit" the data? From the equation, find y, when x = 55.0.

x	11.1	23.2	25.6	37.8	49.6	57.3
y	−1.26	−3.55	−3.85	−5.06	−5.09	−6.35

7. Find the exponential regression equation which best "fits" the data. Also give the correlation coefficient. Plot the equation and the data on the same screen. How well does the equation "fit" the data? From the equation, find y, when x = 4.0.

x	0.8	1.4	1.8	2.5	2.9	3.4	3.8	4.6
y	18.5	28.7	40.1	67.7	91.3	133.0	173.8	328.2

8. Find the exponential regression equation which best "fits" the data. Also give the correlation coefficient. Plot the equation and the data on the same screen. How well does the equation "fit" the data? From the equation, find y when x = 1.00.

x	0.15	0.47	0.56	0.63	0.85	0.92	1.08	1.20
y	3.71	3.65	3.51	3.45	3.36	3.24	3.10	3.02

◆ 9. *(Washington, Exercises 22-5, #9)* The makers of a special blend of coffee found that the demand for the coffee depended on the price charged. The price P per pound and the monthly sales S are shown in the following table. Find the power regression equation which best "fits" the data. S is the independent variable. Also give the correlation coefficient. Plot the equation and the data on the same screen. How well does the equation "fit" the data? From the equation, find P, when the sales are 500,000.

S (thousands)	244	308	419	484	562
P (dollars)	5.62	4.43	3.18	2.83	2.42

◆ 10. *(Washington, Exercises 22-5, #10)* The resonant frequency of an electric circuit containing a 4-μF capacitor was measured as a function of an inductance in the circuit. The following table gives the data obtained. Find the power regression equation which best "fits" the data. Also give the correlation coefficient. Plot the equation and the data on the same screen. How well does the equation "fit" the data? From the equation, find y, when x = 5.00.

L (henrys)	1.1	2.1	4.2	6.2	9.1	12.1
f (hertz)	492	363	251	204	175	156

In exercises 11-16, enter the given data into the calculator. Try the different regression methods available on the calculator and select the equation with the regression coefficient nearest to −1 or to 1 to represent the data. Then, use the equation to answer the questions in the exercise.

11. The resistance in a resistor is measured at various temperatures. The results are summarized in the following table. Resistance is a function of temperature.

Temperature (°C)	4.2	8.0	12.1	15.8	20.3
Resistance (ohms)	0.880	0.918	0.941	0.980	1.014

(a) What is the resistance at 10.0°C?
(b) What is the resistance at 22.0°C?
(c) At what temperature is the resistance 1.000 ohms?
(d) At what temperature is the resistance 1.020 ohms?

12. The pressure, in atmospheres, in a steel cylinder is measured at various temperatures, in degrees Celsius. The results are summarized in the following table. Temperature is the independent variable.

Temperature (°C)	20.5	29.6	39.7	50.6	62.3	70.5
Pressure (atm)	1.22	1.77	2.37	3.03	3.73	4.22

(a) What is the pressure at 45.0°C?
(b) What is the pressure at 10.0°C?
(c) At what temperature is the pressure 2.00 atm.?
(d) At what temperature is the pressure 1.00 atm.?

13. A student measures the resistance in a certain wire as a function of diameter of the wire. The data is summarized in this table.

Diameter (inches)	0.128	0.104	0.080	0.064	0.048
Resistance (ohms)	0.302	0.466	0.775	1.232	2.190

(a) What is the resistance of 0.036 inch diameter wire?
(b) What is the resistance of 0.160 inch diameter wire?
(c) What diameter wire would have a resistance of 1.500 ohms?

14. In a laboratory experiment a student measures the period (in seconds) of a simple pendulum for various lengths (in centimeters) and arrives at the following data. Find the period as a function of length.

Length (cm)	200.0	220.0	240.0	260.0	280.0	300.0	320.0
Period (sec)	21.2	22.3	23.1	24.3	25.0	26.0	26.8

(a) What would the period be for a pendulum of length 250 cm.?
(b) What would the period be for a pendulum of length 350 cm.?
(c) What length would the pendulum have to be to have a period of 10.0 sec.?

15. To measure the atmospheric pressure at different altitudes, a balloon is released containing an instrument which radios back pressure readings at various altitudes. The resulting data is as follows. Pressure is a function of altitude. (Note: For this exercise be careful about rounding constants.)

Altitude (feet)	721.0	3115	6335	8224	10032
Pressure (lb./in^2)	14.25	13.08	11.61	10.81	10.14

(a) The height of Mt. Rainier in Washington State is 14,410 feet. What is the pressure on the top of Mt. Rainier?
(b) According to this data, what is the air pressure at sea level (0 feet)?
(c) At what altitude would the pressure be 5.00 lb./in^2?

16. First class postage rates for United States are listed in the following table for years since 1919.

Year (19--)	19	32	58	63	68	71	74	75	78	81	81	85	88	91	95
Rate (cents)	2	3	4	5	6	8	10	13	15	18	20	22	25	29	32

(a) By hand, graph postage as a function of year on regular graph paper.
(b) If you have semi-logarithm graph paper, graph the rate on the logarithmic scale and the year on the linear scale. If you don't have semi-logarithm graph paper, it will be necessary to first use a calculator to find the logs of the rate values. Then, plot the logs of the rates as the dependent variable and the year as the independent variable.
(c) In some cases, when working with data, it is necessary to disregard some of the data. In this case, in finding an equation, we will disregard the data prior to 1955. Based on the graphs in parts (a) and (b), why should we do this?
(d) Enter the data into your calculator, but use the year 1955 as time (the x variable) zero. Then 1958 will be x = 3, 1963 will be x = 8, 1971 will be x = 16, etc. These correspond to the data points (3, 4), (8, 5), and (16, 8). (For 1981, enter both data points.) Then, based on the coefficient of regression, what is the equation that best "fits" the data.
(e) Graph the equation on the same graph as the data of part (a). How do the graphs compare?
(f) Graph the equation on the same graph as the data of part (b). How do the graphs compare?

For parts (g) through (j), use the equation of part (d).

(g) Calculate the rate for years 1960 and 1980. How much different are these than the actual rate?

(h) Calculate the rate for the present year. How much different is this than the actual rate?

(i) Calculate the rates that could be expected five years from now and ten years from now.

(j) Determine the year that the rate could be expected to be 50 cents.

17. The public debt of the United States is given in the following table for certain years since 1900.

Year (19--)	20	30	40	50	55	60	65
Debt (in billions of dollars)	24.3	16.2	43.0	256.1	272.8	284.1	313.8

Year (19--)	70	75	80	85	90	93
Debt (in billions of dollars)	370.1	533.2	907.7	1823.1	3233.3	4351.2

(a) Graph debt as a function of year on regular graph paper.

(b) If you have semi-logarithm graph paper, graph the debt on the logarithmic scale and the year on the linear scale. If you don't have semi-logarithm graph paper, it will be necessary to first use a calculator to find the logs of the debt values. Then, plot the logs of the debts as the dependent variable and the year as the independent variable.

(c) In some cases, when working with data, it is necessary to disregard some of the data. In this case, when we try to find an equation relating this data, we will disregard the data prior to 1964. Based on the graphs in parts (a) and (b), why should we do this?

(d) Enter the data into your calculator, but use the year 1964 as time (the x variable) zero. Then 1965 will be $x = 1$, 1970 will be $x = 6$, 1975 will be $x = 11$, etc. Then, based on the coefficient of regression, determine which equation gives the best "fit" to the data.

(e) Graph the equation on the same graph as the data of part (a). How do the graphs compare?

(f) Graph the equation on the same graph as the data of part (b). How do the graphs compare?

For parts (g) and (h), use the equation of part (d).

(g) Calculate the debt for years 1970 and 1980. How much different are these than the actual rate?

(h) Calculate the debts that could be expected this year and in the year 2010.

12.3 The Normal Curve

In Section 12.1, we saw that data is often visualized by frequency polygons or histograms. In many cases, particularly with large amounts of "naturally" occurring data, the frequency polygons or histograms appear bell shaped as in Figure 12.5. The highest point on the graph will occur near the mean for the set of data and, as the distance from the mean becomes greater, the height of the graph will approach zero. Such a distribution of data is called a **normal distribution** and the related graph is called the **normal distribution curve**.

If we consider a normal distribution with a mean of zero and a standard deviation of one, we obtain the **standard normal distribution**. The equation of the standard normal distribution curve is

$$y = \frac{1}{\sqrt{2\pi}} e^{-\frac{x^2}{2}} \quad \text{(Equation 1)}$$

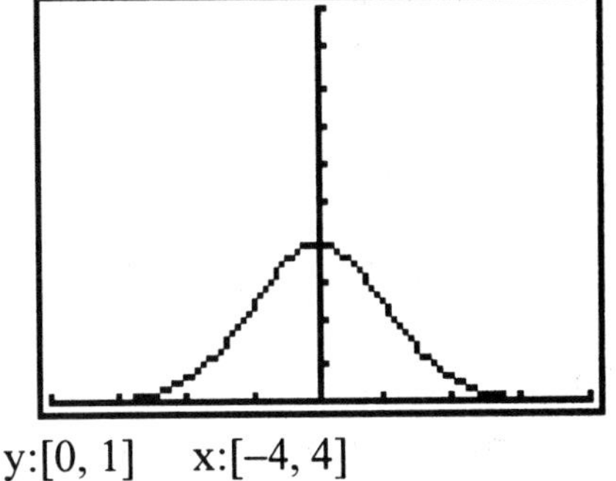

y:[0, 1] x:[–4, 4]

Figure 12.5

The area under the graph of this function is exactly one square unit. The graph is symmetrical with respect to the y-axis with half of the area being to the right of the mean and half being to the left of the mean. The y-intercept is $\frac{1}{\sqrt{2\pi}} e^0$ or about 0.40.

For the standard normal distribution curve the values on the x-axis are actually the number of standard deviations from the mean and are called **z-values**. Positive values of z represent the number of standard deviations above the mean and negative values represent the standard deviations below the mean. Thus, a value of zero is at the mean and a value of one on the x-axis represents a location one standard deviation above the mean. For a set of data that has a normal distribution with a mean of 25 and a standard deviation of 3, the location of 1 on the x-axis ($z = 1$) would represent a data value of the mean plus one standard deviation or $25 + 3 = 28$. Likewise, the location of –2.5 on the x-axis ($z = -2.5$) would represent a data value of the mean less two and one-half standard deviations or $25 - 2.5*3 = 17.5$.

The relationship between a data value, x_1, and the corresponding z-value is given by

$$z = \frac{x_i - \bar{x}}{s} \quad \text{(Equation 2)}$$

where s is the standard deviation and $\bar{x}$ is the mean of the data values.

Of particular interest is the area under the graph of the standard normal distribution curve between two z-values. Since the area under the complete curve (from negative infinity to positive infinity) is one, the area between two z-values represents the likelihood or the probability that data values will fall between these two z-values.

To determine the area between two z-values on the calculator, we will need to write a program for calculating area under the graph. In more advanced mathematics (calculus), the area between the graph, the x-axis, and between two values of x may be found by using something called the definite integral of a function. We will not be concerned at this time with the details of the definite integral but only with the fact that it can be used to find areas. Some graphing calculators have a built in function to evaluate the definite integral and, thus, to find the area under a graph.

To evaluate the area under the graph of the normal distribution function, we first store the standard normal distribution function (Equation 1) for a function name. Then, we enter a short program into the calculator that asks for two z-values and uses the definite integral to evaluate the area between these two z-values. The reader should refer to Section 5 of Chapter 2 in regards to entering programs and using the Input and Disp statements.

Procedure P7. Finding the area under a graph.

1. The desired function should be stored under the function name in the function table. (Procedure C6)
2. Enter the following program into your calculator to find the area between two values B and C:
   (It is assumed that the function is stored under the function name Y_1)

```
PROGRAM:AREA
Disp "ENTER FIRST Z"
Input B
Disp "ENTER SECOND Z"
Input C
fnInt(Y1,X,B,C)→I
Disp "AREA IS:"
Disp I
```

To obtain the definite integral function fnInt:

On the TI-82:	On the TI-85:
Press the key MATH	Press the key CALC
From the menu, select 9:fnInt(	From the menu, select fnInt

The form of the definite integral function is

fnInt(function name, variable, smaller value, larger value)

(The function may also be used without being in a program.)

Example 12.6: Use the program AREA to find the area between $z = 0$ and $z = 5$.

Solution:

1. Store Equation 1 for a function name and use Procedure P7 to find the area between the graph, above the x-axis, and between the z-values of 0 and 5.
2. The area under the standard normal distribution curve between $z = 0$ and $z = 5$ will be given as .4999997072. This indicates that almost all of the area to the right of the mean is contained between $z = 0$ and $z = 5$. Rounding this number to the 4 decimal place accuracy, we obtain 0.5000.

It is possible to add a statement to the program that will store the standard normal distribution function for a function name. To do this, add a line to the program to do the process given in Procedure P7. This is left as an exercise.

Assuming that a set of data has a normal distribution we may use the area under the normal distribution curve to calculate the probability and percent likelihood that certain values of data lie within a given range. If the total number of data items is known, we can use this probability to determine how many of the data items can be expected to lie within the given range.

Example 12.7: Assuming a normal distribution, determine the probability and percent likelihood that a data item lies within 1 standard deviation below the mean and 1.5 standard deviations above the mean.

Solution:

1. We need the area under the standard normal distribution curve between $z = -1$ and $z = 1.5$ as illustrated by the shaded area in Figure 12.6. This area gives the fraction of the total area (one square unit) and thus the probability that the data lies between 1 standard deviations below the mean and 1.5 standard deviations above the mean.
2. Use the calculator and the Procedure P7 to find the required area (See Example 12.6). Use –1 for the first z-value and 1.5 for the second z-value and round the answer to 4 decimal places. The area is .7744 or 77.44% of the total area.

   Since the area is .7744, the probability that a data item lies in the specified range is .7744.

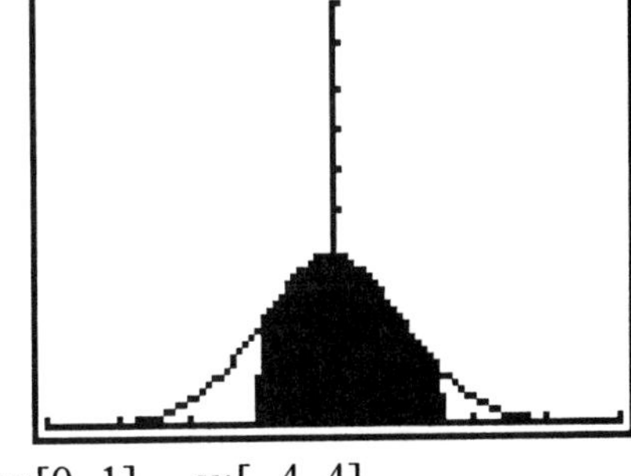

y:[0, 1] x:[–4, 4]

Figure 12.6

The percent likelihood that a data item lies within this range is 77.44%

Example 12.8: A store receives a shipment of 5000 light bulbs with a mean life span of 1000 hours and standard deviation of 25 hours. Assume the data gives a normal distribution. Determine the percent likelihood that
(a) a bulb will last between 970 hours and 1040 hours.
(b) a bulb will last at least 1050 hours.

Solution:

1. We first use Equation 2 to determine the z-values that correspond to the times. For this data $\overline{x} = 1000$ and $s = 25$. Thus,

$$\text{For } x = 970,\ z = \frac{970 - 1000}{25} = -2.5$$

$$\text{For } x = 1040,\ z = \frac{1040 - 1000}{25} = 1.6$$

$$\text{For } x = 1050,\ z = \frac{1050 - 1000}{25} = 2.0$$

2. Use the Procedure P7 to find areas under the normal curve as in Example 12.6.
   (a) The area between $z = -1.2$ and $z = 1.6$, is 0.8300.
   (b) In order to determine if a bulb will last longer than 1050 hours, we need the area to the right of $z = 2.0$. Since the area past $z = 5$ is negligible (to 4 decimal places) find the area between $z = 2$ and $z = 5$.
3. (a) Since the area between $z = -1.2$ and 1.6 is 0.8300, 83.00% of the values lie between $x = 970$ and $x = 1040$. The number of bulbs with a life span between 970 and 1040 hours is 83.00% of 5000 or 4150 bulbs.
   (b) The area above $z = 2$ is 0.0227 and 2.27% of the values lie above $x = 1050$. The number of bulbs with a life span of at least 1050 hours is 2.27% of 5000 or 113.5 bulbs. Since we cannot have a fractional bulb, the number of bulbs that can be expected to last 1050 hours is 113.

Many times it is important to know what values of data fall within a certain percentage range. This is called a **confidence interval**. To determine the 80% confidence interval for a set of data, we determine two data values, equally spaced on either side of the mean, such that 80% of the data lie between these two values.

Example 12.9: Determine the z-value (to two decimal places) such that 10% of the total area is to the right of this z-value as in Figure 12.7.

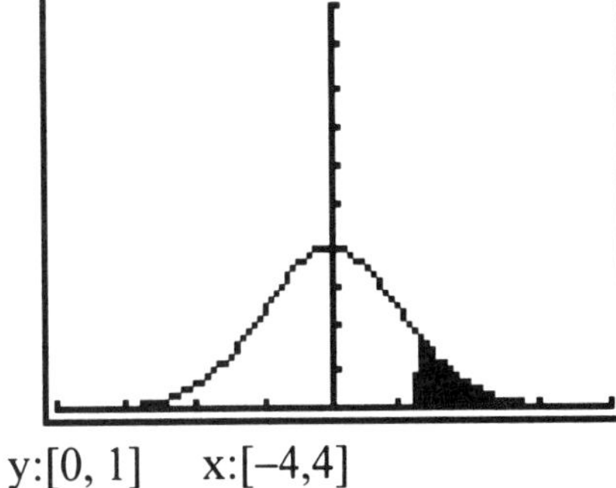

y:[0, 1] x:[−4,4]

Figure 12.7

Solution: We are not able to calculate the z-values directly, thus, we to use a trial and error method with Procedure P7.

1. Since 10% of the area is to be to the right of the z-value and we know that 50% of the area is to the right of the mean, 40% of the area will be between the mean and this z-value. Thus, we desire a value of z such that the area is between z = 0 and this z-value is 0.4000.
2. Start by finding the area between z = 0 and z = 1 which is .3413. Thus, the z-value we seek must be greater than 1.
3. Next, find the area between z = 0 and z = 2. This gives an area of .4772. The z-value we seek must lie between 1 and 2.
4. Then, find the area between z = 0 and z = 1.5. The area is .4332 and the z-value must lie between 1 and 1.5. We continue this process of narrowing the interval until we obtain the required z-value. With the first z value equal to zero, we obtain on successive calculations

   For second z = 1.25, area = 0.3943
   For second z = 1.30, area = 0.4032
   For second z = 1.28, area = 0.3997
   For second z = 1.29, area = 0.4015

   Since the area for 1.28 is closer to 0.4000 than that for 1.29, we conclude that, to two decimal places, the required z-value is 1.28. Thus, under the normal distribution curve, 10% of the total area is to the right of z = 1.28.

Example 12.10: For the shipment of light bulbs of Example 12.7, with a mean life span of 1000 hours and a standard deviation of 25 hours, determine the 80% confidence interval.

Solution: To determine the 80% confidence interval, we determine the life span values between which 80% of the data will lie. Since these are equally spaced on either side of the mean, 40% will lie above the mean and 40% below the mean.

1. We first determine the z-value such that 40% of the area under the normal distribution curve will be above the mean. Using the procedure of Example 12.8, we find this value to be 1.28.
2. The z-value of 1.28 indicates 1.28 standard deviations above the mean or (1.28)(25) = 32 hours above the mean is the upper limit for the 80% confidence level. Since, the graph is symmetrical, 32 hours below the mean will be the lower limit.
3. Thus, an 80% confidence interval will be those values within 32 hours of the mean 80% of the light bulbs can be expected to have a life span of between 968 and 1032 hours.

Exercise 12.3

1. Use Procedure P7 to find the area under the normal distribution curve between
   (a) $z = -1$ and $z = 1$
   (b) $z = 0$ and $z = 2$
   (c) $z = 3$ and $z = 5$
2. Use Procedure P7 to find the area under the normal distribution curve for
   (a) $z = -2$ and $z = 2$
   (b) $z = -4$ and $z = 0$
   (c) $z = -5$ and $z = 5$
3. Add a line to the program in Procedure P7 to store the standard normal distribution function for a function name in the function table. Then, find the area under the normal distribution curve between $z = -1$ and $z = 1$.
4. Use the program of Exercise 3 to find the area under the normal distribution curve between
   (a) $z = -1.5$ and $z = 2.5$
   (b) $z = -2.0$ and $z = 1.4$
   (c) $z = 1$ and $z = 2.3$
5. For data which has a normal distribution, find the probability and percent likelihood that the data
   (a) Lies between the mean and 2.2 standard deviations above the mean.
   (b) Lies between 1.45 standard deviations below the mean and 2.25 standard deviations above the mean.
   (c) Lies to the left of 1.82 standard deviations below the mean.
   (d) Lies within 1.80 standard deviations of the mean.
6. For data which has a normal distribution, find the probability and percent likelihood that the data

(a) Lies between the mean and 1.3 standard deviations below the mean.
(b) Lies between 2.22 standard deviations below the mean and 1.25 standard deviations above the mean.
(c) Lies to the right of 2.20 standard deviations above the mean.
(d) Lies more than 1.75 standard deviations from the mean.

7. A set of data has a mean of 73.5 with a standard deviation of 2.5. Find the percent likelihood that a data value
   (a) Lies between 68.5 and 76.0.
   (b) Is greater than 74.7.
   (c) Is greater than 69.2 and less than 74.2.
   (d) At least 75.2
8. A set of data has a mean of 0.712 with a standard deviation of 0.012. Find the percent likelihood that a data value
   (a) Lies between 0.700 and 0.725.
   (b) Is greater than 0.730.
   (c) Is greater than 0.715 and less than 0.718.
   (d) At least 0.705.
9. A sample of half inch bolts shows a mean diameter of 0.5030 inches with a standard deviation of 0.0025 inches. Find the probability that
   (a) A bolt will exceed 0.5050 inches.
   (b) A bolt will lie between 0.5000 and 0.5040 inches.
   (c) A bolt will be less than 0.4995 inches.
   (d) A bolt will exceed 0.5030 inches.
10. A sample of 20 ounce cereal boxes has a mean weight of 20.35 ounces with a standard deviation of 0.25 ounces. Find the probability that
    (a) A box of cereal will exceed 20.50 ounces.
    (b) A box of cereal lie between 20.00 and 21.00 ounces.
    (c) A box of cereal will be less than 20.00 ounces.
    (d) A box of cereal will exceed 20.35 ounces.
11. A sample of resistors shows a mean resistance of 1.205 ohms with standard deviation of 0.015 ohms. Out of 2000 resistors, determine how many will
    (a) Lie between 1.195 ohms and 1.210 ohms.
    (b) Have a resistance greater than 1.212 ohms.
    (c) Have a resistance less than 1.184 ohms.
    (d) Have a resistance less than 1.196 ohms or greater than 1.214 ohms.
12. A sample of toner cartridges for a printer has a mean life span of 15,500 copies with standard deviation of 320 copies. Out of 500 cartridges, determine how many can be expected to have a life span
    (a) Between 15,000 and 16000 copies.
    (b) Greater than 16,600 copies.
    (c) Less than 14,500 copies.
    (d) Less than 14,800 copies or greater than 15,800 copies.

13. Use the data of Exercise 15 of Section 12.1 and determine how many light bulbs out of 1200 can be expected to have a life span of
    (a) Between 1180 and 1220 hours.
    (b) Greater than 1230 hours.
    (c) Less than 1175 hours.
    (d) Greater than 1175 hours.
14. Use the data of Exercise 12 of Section 12.1 and determine the likelihood of a car stopping
    (a) In less that 162 feet.
    (b) In more than 177 feet.
    (c) In more than 182 feet.
    (d) Between 162 and 177 feet.
15. For a normal distribution, find the z-values such that
    (a) 25% of the total area is to the right of the z-value.
    (b) 5% of the total area is to the left of the z-value.
    (c) 90% of the total area is around the mean (45% above and 45% below)
    (d) 99% of the total area is around the mean.
16. For a normal distribution, find the z-values such that
    (a) 15% of the total area is to the right of the z-value.
    (b) 8% of the total area is to the left of the z-value.
    (c) 95% of the total area is around the mean (45% above and 45% below)
    (d) 98% of the total area is around the mean.
17. For the data in Exercise 9, find
    (a) A 90% confidence interval.
    (b) A 95% confidence interval.
    (c) A 99% confidence interval.
18. For the data in Exercise 10, find
    (a) A 90% confidence interval.
    (b) A 95% confidence interval.
    (c) A 99% confidence interval.
19. For the data in Exercise 11, find
    (a) A 80% confidence interval.
    (b) A 90% confidence interval.
    (c) A 97.5% confidence interval.
20. For the data in Exercise 12, find
    (a) A 80% confidence interval.
    (b) A 90% confidence interval.
    (c) A 97.5% confidence interval.

Chapter 13

Limits and Derivatives

13.1 Limits
(Washington, Section 23-1)

In this chapter we open the door to an area of mathematics called calculus. Calculus is very important from many standpoints in today's world. To understand calculus, it is first necessary to understand the concept of limit. With limits we are concerned with the value that a function approaches as the independent variable comes close to a given value.

For example, the function $f(x) = x^2$ has a value of $f(4) = 16$ at $x = 4$. We can also say that the limit of f(x) as x approaches 4 is 16. This last statement follows because as we let take values of x that get closer to 4 -- values that are either greater than or less than 4 -- the corresponding values of the function get closer to 16. We write this as $\lim_{x \to 4} x^2 = 16$.

In general, the **limit L of a function** f(x) is the value *L* that the function approaches as x approaches a number *a*. We write this as $\lim_{x \to a} f(x) = L$ and read it: the limit of f(x) as x approaches *a* is *L*. In finding a limit, we are not concerned with the value of the function exactly at *a*, or even if the function exists at *a*, but we are concerned with what value the function approaches as x gets very close to *a*.

The limits to a function as x approaches a given value may be estimated numerically by evaluating the function at several x-values where each value is closer to the given value. For example, if we wish to find $\lim_{x \to 1} f(x)$, we might first evaluate the function at values greater than 1 such as 1.5, 1.2, 1.1, 1.05, 1.01, 1.001, etc. This would give what is called the **right hand limit.** For the right hand limit we use the notation $\lim_{x \to 1^+} f(x)$. We could also evaluate the function at values less than 1 such as 0.5, 0.8, 0.9, 0.95, 0.99, 0.999, etc. This gives what is called the **left hand limit** and is indicated by the notation $\lim_{x \to 1^-} f(x)$. **Only when the right hand limit and the left hand limit both exist and are equal does the limit itself exist.**

It is important to note that estimating a limit numerically does not mathematically prove that the limit actually exists nor that the value is the number we estimate. In this material, we shall only consider the numerical estimation of limits.

The estimation of a limit using the graphing calculator requires the evaluation of the function at several values of x. For methods of evaluating functions, refer to Section 1 of Chapter 2 and to Procedure C11(using a list may be particularly helpful). It may also be helpful to write a short program as Example 2.11. Another method that offers more insight is to evaluate the function while the graph is on the screen by using the trace function or as in Procedure G13.

♦ Example 13.1: *(Washington, Section 23-1, Examples 7 and 8)* Estimate $\lim_{x \to 2} \frac{2x^2 - 3x - 2}{x - 2}$ by graphing the function and evaluating both the right hand and left hand limits.

Solution:

1. Graph this function using the standard viewing rectangle. A straight line is obtained and there appears to be a point on the graph for $x = 2$. When we use Procedure G13 to evaluate the function at $x = 2$ no value appears for y. (What happens when we try substituting $x = 2$ directly into the function?)
2. Select the decimal viewing rectangle (see Procedure G6) and change the viewing rectangle so that Ymin = 0 and Ymax = 7. The graph is shown in Figure 13.1 has a hole appearing at $x = 2$ (since the function is undefined at $x = 2$).
3. Use the trace function and notice the y values as x approaches 2 from the left side. What number is approached by the y-values? Repeat the process as x approaches 2 from the right hand side. What number is approached by the y-values? From this experiment we conclude that the left hand limit and the right hand limit are both 5 and that $\lim_{x \to 2} \frac{2x^2 - 3x - 2}{x - 2} = 5$. We could also zoom in to obtain x-values that are closer to 2 to help verify the limit.

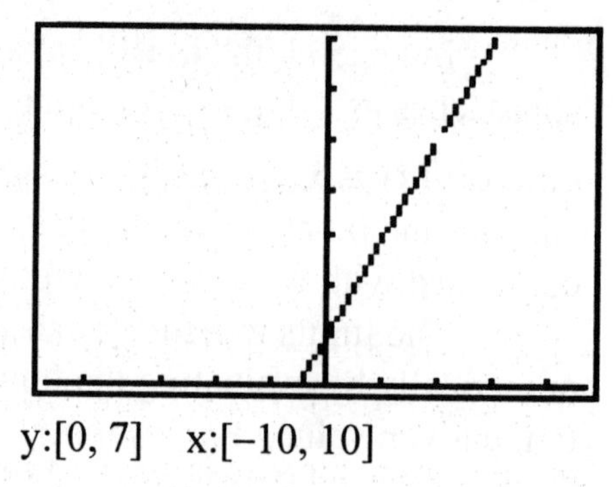

y:[0, 7] x:[−10, 10]

Figure 13.1

4. Next, evaluate the function at several values of x with the graph on the screen. To estimate the right hand limit, use Procedure G13 and evaluate the function at x = 1.9, 1.95, 1.99, 1.999, and 1.9999. Then estimate the left hand limit by evaluating the function at x = 2.1, 2.05, 2.01, 2.001, 2.0001. Round all values to 4 decimal places to obtain the following values:

x	1.9	1.95	1.99	1.999	1.9999
y	4.8	4.9	4.98	4.998	4.9998

x	1.9	1.95	1.99	1.999	1.9999
y	4.8	4.9	4.98	4.998	4.9998

5. Based on this table of values we estimate that $\lim\limits_{x\to 2^-} \dfrac{2x^2-3x-2}{x-2} = 5.0$ and that $\lim\limits_{x\to 2^+} \dfrac{2x^2-3x-2}{x-2} = 5.0$.
   Since both the right hand and left hand limits are the same we conclude that $\lim\limits_{x\to 2} \dfrac{2x^2-3x-2}{x-2} = 5.0$.

In some cases, the y-values may become very large positive values or very large negative values as x approaches a given value from either the right hand or left hand side. If the y-values approach a very large positive number we say the limit is $+\infty$ and if they approach a very large negative number we say the limit is $-\infty$. In this case, the limit itself does not exist.

In other cases, a function may approach a value as x becomes a very large positive number (approaches positive infinity, $+\infty$) or as x becomes a very large negative number (approaches negative infinity, $-\infty$). For Example, the function $\dfrac{1}{x}$ approaches zero as x approaches ∞ or as x approaches $-\infty$. To see what happens, as x becomes large positive values evaluate the function at numbers such as 100, 1000, 10000, 100000, etc., or as x becomes large negative values, at numbers such as −100, −1000, −10000, −100000, etc. If the y-values approach a real number L, we say that a limit exists and is equal to the value L. We write $\lim\limits_{x\to +\infty} f(x) = L$ for a limit as x approaches a large positive number and $\lim\limits_{x\to -\infty} f(x) = L$ for a limit as x approaches a large negative number.

Example 13.2: Estimate $\lim\limits_{x\to +\infty} \dfrac{2x^2-1}{x^2+2}$ and $\lim\limits_{x\to -\infty} \dfrac{2x^2-1}{x^2+2}$.

Solution: 1. Graph the function $y = \dfrac{2x^2-1}{x^2+2}$ on the standard viewing rectangle.
The graph is shown in Figure 13.2. Use the trace function and let x increase in value. The y-values become larger and get closer to the number 2 but never exceeds 2. If we use the trace function and let x

become large negative numbers, we notice that again the y-values become larger and get closer to the number 2 but never exceeds 2. We estimate the limit in each case as being 2 and write

$\lim_{x \to +\infty} \frac{2x^2 - 1}{x^2 + 2} = 2$ and

$\lim_{x \to -\infty} \frac{2x^2 - 1}{x^2 + 2} = 2$. In some cases it would be best to take a larger Xmax and a smaller Xmin.

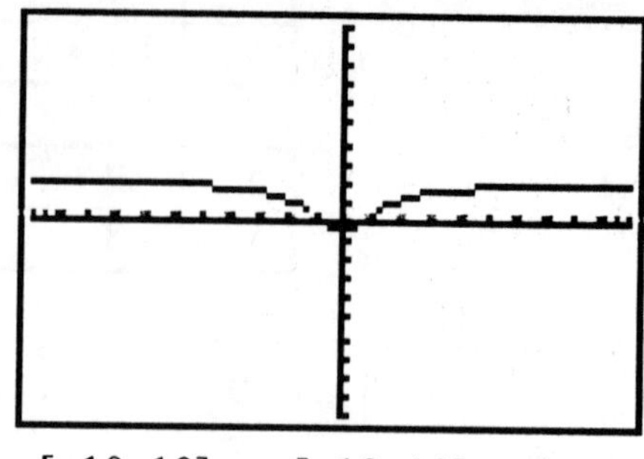

y:[−10, 10]   x:[−10, 10]

Figure 13.2

2. The limits may also be estimated numerically by evaluating the function using appropriate values of x to estimate the limit:

x	5	10	100	1000
y	1.814814	1.950980	1.999500	1.999995

B

a x	−5	−10	−100	−1000
s y	1.814814	1.950980	1.999500	1.999995

e

d on this table we estimate $\lim_{x \to +\infty} \frac{2x^2 - 1}{x^2 + 2} = 2$ and $\lim_{x \to -\infty} \frac{2x^2 - 1}{x^2 + 2} = 2$

Exercises 13.1

In Exercises 1-12, find the right hand limit, the left hand limit and the limit itself if it exists. Indicate when a limit doesn't exist. Use both the graphs and numerical evaluation of the function to estimate the limits. Sketch the graph of the function. When finding limits, indicate the values of x used and the values of the function.

1. $\lim_{x \to 2} (x^2 - 3x)$
2. $\lim_{x \to -3} (4 - x^2)$
3. $\lim_{x \to 2} \frac{x^2 - 4}{x - 2}$
4. $\lim_{x \to 9} \frac{x^2 - 9x}{x - 9}$

5. $\lim_{x \to \sqrt{3}} \frac{x^4 - 6x^2 + 9}{x^2 - 3}$
6. $\lim_{x \to 0} \frac{4 - \sqrt{x - 16}}{x}$
7. $\lim_{x \to 0} \frac{(2 + x)^2 - 3(2 + x) + 2}{x}$
8. $\lim_{x \to 0} \frac{-16(1 + x)^2 + 16}{x}$
9. $\lim_{x \to 1.5} \frac{1 - x}{x - 1.5}$
10. $\lim_{x \to 3} \frac{x - 2}{x - 3}$

◆ 11. *(Washington, Exercises 23-1, #54)* $\lim_{x \to 0} \frac{\sin x}{x}$ [Place calculator in radian mode]

12. $\lim_{x \to 0} 4^{\frac{1}{x}}$

In Exercises 13-18 estimate the limit both from the graph of the function and by evaluating the function at several appropriate values of x. Indicate the values of x used and the corresponding values of the function. Estimate limits to three significant digits.

13. $\lim_{x \to \infty} \frac{3x + 3}{4x - 1}$
14. $\lim_{x \to \infty} \frac{4.7x^2 - 3x + 2}{2x^2 + 4x}$
15. $\lim_{x \to \infty} \frac{32 - 12.64x^3}{4x^3 + 24x}$
16. $\lim_{x \to -\infty} \frac{125 - 75x^2 + 61x^3}{94x^3 - 72x}$

◆ 17. *(Washington, Exercises 23-1, #53)* $\lim_{x \to \infty} (1 + x)^{\frac{1}{x}}$

◆ 18. *(Washington, Exercises 23-1, #52)* A 5.25 ohm resistor and a variable resistor of resistance R are placed in parallel. The expression for the resulting resistance R_T is given by $R_T = \frac{5.25R}{5.25 + R}$. To estimate the value of the resulting resistance as R becomes large, find $\lim_{R \to \infty} R_T$.

19. The average rate of change of y with respect to x for the function $y = x^2 - 2$ is given by the expression $\frac{\Delta y}{\Delta x} = \frac{(x + \Delta x)^2 - 2 - (x^2 - 2)}{\Delta x}$. To find the instantaneous

rate of change at $x = 2$, substitute 2 for x and determine the limit as Δx approaches 0 of $\frac{(2+\Delta x)^2 - 2 - (4-2)}{\Delta x} = \frac{(2+\Delta x)^2 - 4}{\Delta x}$. Do this by graphing the equivalent function $y = \frac{(2+x)^2 - 4}{x}$ and finding the limit as x approaches zero. Estimate the limit both from the graph and by evaluating the expression.

20. If the distance y, in miles, a car travels as a function of time t, in minutes, is given by the expression $y = 30t + 0.5t^2$, then the average rate of change is given by the expression $\frac{\Delta y}{\Delta t} = \frac{30(t+\Delta t) + 0.5(t+\Delta t)^2 - (30t + 0.5t^2)}{\Delta t}$. To find the instantaneous rate of velocity at $t = 5$ minutes, substitute 5 for t, simplify, and find the limit as Δt approaches 0. (See Exercise 19.) Estimate the limit from the graph and by evaluating the expression.

13.2 The Derivative
(Washington, Sections 23-3 and 23-4)

Consider the ratio $\frac{\Delta y}{\Delta x}$ for the function $y = f(x)$ where $\Delta y = f(x + \Delta x) - f(x)$. As Δx approach zero, we find that this ratio $\frac{\Delta y}{\Delta x}$ approaches a fixed value (see Exercises 19 and 20 of Exercise Set 13.1). The limit (as $\Delta x \to 0$) of this ratio is a very important quantity in mathematics and is called the derivative of the function. More formally we define the **derivative** of a function as

$$\lim_{\Delta x \to 0} \frac{\Delta y}{\Delta x} \quad \text{or} \quad \lim_{\Delta x \to 0} \frac{f(x+\Delta x) - f(x)}{\Delta x} \qquad \text{[Equation 1]}$$

Another equivalent definition of the derivative is

$$\lim_{\Delta x \to 0} \frac{f(x+\Delta x) - f(x-\Delta x)}{2\,\Delta x} \qquad \text{[Equation 2]}$$

To approximate the derivative of the function at $x = a$, either Equation 1 or Equation 2 may be evaluated when $x = a$ for values of Δx as Δx approaches zero. The value obtained is called the numerical derivative. Some graphing calculators determine the value of the numerical derivative by a similar formula. Since different notations may be used to represent the numerical derivative, we will use nDeriv.

Procedure C19. To determine the numerical derivative from a function.

On the TI-82:
1. Press the key MATH
2. Select 8:nDeriv(
   The form is
   nDeriv(y, variable, value)

On the TI-85:
1. Press the key CALC
2. From the menu, select nDer
   The form is
   nDer(y, variable, value)

where y is the function itself or the name of the function, variable is the variable used in the function, and value is the number at which we wish to determine the derivative.

TI-82: Example: nDeriv(Y_1,x,2)
The precision to which this is calculated may be changed by indicating the precision as a fourth value inside the parentheses. The value used otherwise is 0.001. (This value determines the Δx.)

TI-85: Example: nDer(y1,x,2)
The precision to which this is calculated may be changed by pressing the key TOLER and changing the value for δ (this determines the Δx).
Note: The TI-85 also has der1 which determines the numerical first derivative as exactly as is possible and der2 which determines the numerical second derivative as exactly as possible.

Example 13.3: Find the numerical derivative of $y = x^3 - 2x^2 + 4$ at $x = 1$ and at $x = \frac{4}{3}$. Round answers to two decimal places.

Solution: Use Procedure C19 to determine the numerical derivative of the function with x as the variable. Evaluate first with 1 substituted for value then with $\frac{4}{3}$ substituted for value. The derivative at $x = 1$ is $-.999999$ or -1.00. The value at $x = \frac{4}{3}$ appears similar to 1E−6 (or 0.000001) which rounds off to 0.00.

The value of the derivative of a function at a given x-value gives the instantaneous rate of change of y with respect to x. This instantaneous rate of change may represent the slope of a tangent line on a graph, the instantaneous velocity of a moving object, or the acceleration of a moving object.

In Example 13.3, the derivative of $y = x^3 - 2x^2 + 4$ at $x = 1$ is -1. This means that the instantaneous rate of change of y with respect to x at $x = 1$ is -1 and implies that, at $x = 1$, the y changes -1 unit for a one unit change in x. This also indicates that, at $x = 1$, the slope of the tangent line is -1 and that the slope of the tangent line at $x = \frac{4}{3}$ is 0. (What does this tell us about the tangent line at $x = \frac{4}{3}$?)

The following short program may be helpful in determining values for a numerical derivative of a function at several values of x:

On the TI-82:

```
PROGRAM:DERIV
Disp "ENTER X VALUE"
Input A
nDeriv(Y1,X,A)→Y
Disp Y
```

On the TI-85:

```
PROGRAM :DERIV
Disp "Enter x-value"
Input A
nDer(y1, x, A)→Y
Disp Y
```

When using this program, the function of interest must be stored for the name Y_1. The derivative may be determined for several values of x by repeatedly executing the program by pressing ENTER before pressing any other key. The numerical derivative function is obtained as in Procedure C19 while in programming mode and the arrow → is obtained by pressing the STO▷ key.

Example 13.4: A baseball is thrown vertically at 90 ft/s. The height s, in feet, of the ball at time t, in seconds, is given by the equation $s = 90t - 16t^2$. Find height of the ball and the instantaneous velocity of the ball when $t = 1.0$, 2.0, 2.5, 3.0, and 4.0 seconds. What is the significance of the positive and negative values of the derivative (compare to s-values)?

Solution:

1. In this case, the instantaneous velocity is the instantaneous rate of change of s with respect to t. This instantaneous rate of change may be found by determining the value of the numerical derivative at the given values of t.
2. Evaluating the numerical derivative of the function $s = 90t - 16t^2$, we obtain

t	1.0	2.0	2.5	3.0	4.0
height	74	116	125	126	104
numerical derivative	58	26	10	−6.0	−38

3. From these values we conclude that when the derivative is positive, the height is increasing and when the derivative is negative, the height is decreasing. As the ball approaches its highest point, the value of the derivative approaches zero.

In Example 13.4, we see that when the derivative is positive the dependent variable is increasing and when the derivative is negative the dependent variable is decreasing. This is true of all functions and provides a method of determining when a function is increasing in value or decreasing in value.

Exercise 13.2

In Exercises 1-8, find the numerical derivative of each function for x = –2.0, –1.0, 0.0, 1.0, and 2.0. Compare the value of the function and its derivative at each x-value and determine if there is a relationship between the given function and its derivative.

1. $y = 2x$
2. $y = 5$
3. $y = x^2$
4. $y = x^3$
5. $y = x^4$
6. $y = x^5$
7. $y = x^{-1}$
8. $y = x^{-2}$

In Exercises 9-14, find the numerical derivative at the given values of x and compare to the graph of the function at the given x-values. Determine the values of x for which y is increasing in value and the values of x for which y is decreasing in value.

9. $y = x^2 - 2x$ at x = –1.0, 0.5, 1.0, 1.5, and 2.0.
10. $y = 2 - x - x^2$ at x = –2.0, –0.5, 0.0, 1.0, and 2.0.
11. $y = 12x - x^3$ at x = –4.0, –2.0, 0.0, 2.0, and 3.0.
12. $y = 2x^3 + 6x^2$ at x = –3.0, –2.0, –1.0, 0.0, and 2.0.
13. $y = \dfrac{1}{1+x^2}$ at x = –2.0, –0.5, 0.0, 0.5, and 2.0.
14. $y = \dfrac{1}{(1-x)^2}$ at x = –1.0, 0.0, 1.0, and 2.0.

In Exercises 15-18, find several values of the numerical derivative of the function between x = –5.0 and 5.0 and compare to the graph of the function. Locate values of x at which the function has horizontal tangent lines.

15. $y = 2x^2 - 10x$
16. $y = 9x - x^2$
17. $y = 5x^3 - 9$
18. $y = 6x - x^3$

In Exercises 19-22, find the value of the distance s, in feet, and the instantaneous velocity for the given values of time t, in seconds. Use the numerical derivative to find the instantaneous velocity at the given distance. Note the relationship between the numerical derivative and whether the values of s are increasing or decreasing.

19. $s = 5t - 10$ for $t = 0.0, 1.5, 3.0, 5.5$, and 8.0
20. $s = 24 + 7t$ for $t = 0.0, 2.5, 5.0, 7.5$, and 10.0
21. $s = 76t - 16t^2$ for $t = 1.00, 2.25, 2.375, 2.40$, and 2.50.
22. $s = 135 + 125t - 16t^2$ for $t = 2.50, 3.50, 4.00, 4.50$, and 5.00.

In Exercises 23-26, use the numerical derivative to find the instantaneous rate of change.

23. The cost of producing x parts in a machine shop is given by $C = 4500 - 100x + x^2$. Find the instantaneous rate of change of cost relative to the number of parts produced when $x = 35, 45, 50, 55$, and 65. What is the significance of positive and negative value for the rate of change?
24. A ball bearing is in the shape of a sphere whose volume is given by $V = \frac{4}{3}\pi r^3$, where r is the radius in centimeters. As it is heated it expands. Find the rate of change of volume with respect to the radius when the radius is 6.010 cm, 6.020 cm, and 6.110 cm.
25. ♦ *(Washington, Exercises 23-4, #25)* The total power P, in watts, transmitted by an AM radio station is given by $P = 500 + 250x^2$ where x is the modulation index. Find the instantaneous rate of change of P with respect to x for $x = 0.88$, $x = 0.90$, $x = 0.92$, and $x = 0.94$.
26. ♦ *(Washington, Exercises 23-4, #22)* A load L, in Newton's, is distributed along a beam 10 meters long such that $L = 5x - 0.5x^2$ where x is the distance from one end of the beam. Find the rate of change of L with respect to x when $x = 2.0$ m, 4.5 m, 5.0 m, 5.8 m, and 8.0 m.

13.3 Applications of the Derivative
(Washington, Sections 24-1, 24-2, and 24-5)

In this section we study the use of the graphing calculator in applications related to the graph of a function. The derivative is used to find the equation of a tangent line and graphical means are used to investigate the relationship between the derivative and the function. We are already aware that the derivative of a function evaluated at a point gives the slope of the tangent line at that point. From analytic geometry, we know that we can find the equation of any line if we know the slope, m, of the line and a point, (x_1, y_1), on the line by using the standard form

$$y - y_1 = m(x - x_1). \qquad \text{[Equation 3]}$$

To find the equation of a tangent line using a graphing calculator, use the following procedure:

1. Find the numerical derivative for a given x-value (Procedure C19). This gives the slope of the tangent line.
2. Find the y-coordinate for the given x-value (Procedure G13).
3. Substitute the values into Equation 3 and simplify.

After finding the equation of a tangent line it is interesting to see the tangent line on the same graph as the function. We could just graph the line that we obtain from calculations or we can use the function that is built into many calculators to graph the tangent line.

Procedure G25. Drawing a tangent line at a point.

1. Store the function for a function name and graph the function.
2. Draw the tangent line.

On the TI-82:

(a) While on the calculations screen, press the key DRAW.

(b) Select: 5:Tangent(
The form is
Tangent(y, value)
Where y is the function or function name and value is the x-value.
Example: Tangent(Y_1,2)

(c) Press ENTER

On the TI-85:

(a) With the graph on the screen select DRAW from the GRAPH menu (after pressing MORE)

(b) From the secondary menu, select TanLn (after pressing MORE twice). The form is
TanLn(y, value).
where y is the function or function name and value is the x-value.
Example: TanLn(y1,3)

The tangent line may also be drawn by:

(a) With the graph on the screen and the cursor at the correct point, press DRAW.

(b) Select 5:Tangent(and press ENTER to return to graph.

(c) Press ENTER a second time. The tangent line will appear on the graph and the slope of the line will be given at the bottom of the screen.

The tangent line may also be drawn by:

(a) With the graph and the GRAPH menu on the screen, select MATH (after pressing MORE).

(b) From the secondary menu, select TANLN (after pressing MORE twice).

(c) Move the cursor to the desired point and press ENTER. The tangent line will be drawn on the graph and the it slope given at the bottom of the screen.

Example 13.5: Find the equation of the tangent line to the graph of $y = 9x - 3x^2$ at $x = 2.0$ and use the equation of the tangent line to find the value for y when $x = 2.2$.

Solution:

1. Store the function $9x - 3x^2$ for a function name and graph the function.
2. Determine the coordinates of the point and the slope for $x = 2.0$. The y-coordinate of the point is 6.0 and the slope is 1.5. Use Procedure G25 to graph the tangent line. The graph of the function and the tangent line is in Figure 13.3.
3. From the slope-point form for the equation of a line we have

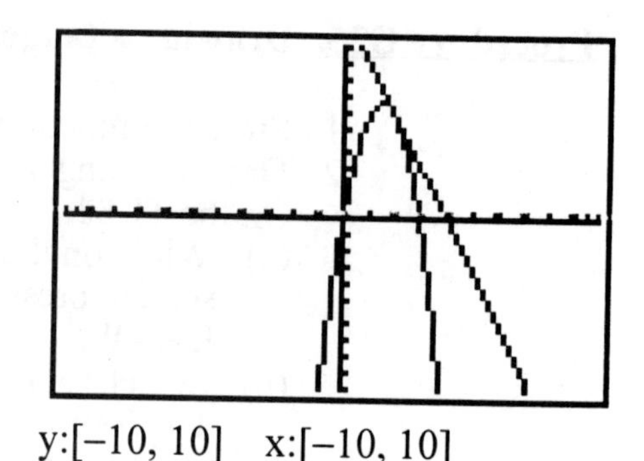

y:[−10, 10] x:[−10, 10]

Figure 13.3

$$y - 6.0 = 1.5\,(x - 2.0)$$
$$\text{or } y - 6.0 = 1.5x - 3.0$$
$$\text{or } y = 1.5x + 3.0$$

4. At $x = 2.2$, $y = 1.5(2.2) + 3.0 = 6.3$.

The derivative of a function and the slope of the tangent line may also help us analyze the graph of a function. We next investigate how the slope of the tangent line is related to the graph of a function.

Procedure G26. Determining the value of the derivative from points on a graph.

On the TI-82:

1. With a graph on the screen, press the key CALC.
2. Select 6:dy/dx.
3. Move the cursor to the desired point and press ENTER. The value of the numerical derivative appears on the bottom of the screen.

On the TI-85:

1. With the and the graph menu on the screen, press the key MATH (after pressing MORE).
2. From the secondary menu select dy/dx.
3. Move the cursor to the desired point and press ENTER. The value of the numerical derivative appears at the bottom of the screen.

Example 13.4: Graph the function $y = x^2 - 2x$ on the standard viewing rectangle, then change to the decimal viewing rectangle. Find several values of the derivative for x less than one and also for x greater than one. What is true about the derivative for x less than one? For x greater than one? Where is the function decreasing and where is it increasing? What is true about the point where $x = 1$?

Solution:

1. Graph the function on the standard viewing rectangle and then change to the decimal viewing rectangle. Evaluate the derivative at $x = -1.0$, $x = 0.0$, $x = 2.0$, and $x = 3.0$ (Procedure C19). We obtain the following values:

x	−1.0	0.0	2.0	3.0
dy/dx	−4.0	−2.0	2.0	4.0

2. Notice that for x-values less than one, the derivative is negative and the graph is decreasing and for x-values greater than one, the derivative is positive and the graph is increasing. At $x = 1$, the derivative is zero and at $x = 1$, there is a local minimum point.

It is also instructive to graph the derivative of a function on the same graph as the function. We can do this even if we do not know the derivative of the function, by first entering the function under one name (Y_1) and then entering the numerical derivative under a second function name (Y_2). In the numerical derivative, use the variable x instead of a constant for value (nDeriv(y,x,x))..

Example 13.4: Graph the function $y = x^2 - 6x + 5$ and the numerical derivative on the same graph. What is true about the function when the graph of the derivative is negative (lies below the x-axis)? What is true about the function when the graph of the derivative is positive (lies above the x-axis)? What is true about the function at the x-intercept of the graph of the derivative?

Solution:

1. Enter and graph the function and the numerical derivative of the function on the standard viewing rectangle. To graph the numerical derivative (assuming that the function has been entered under the name Y_1), enter nDeriv(Y_1,x,x) under a second function name. The graph is in Figure 13.4.

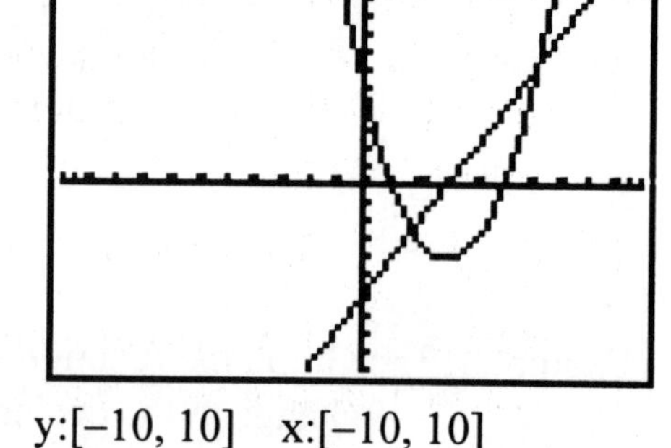

y:[−10, 10] x:[−10, 10]

Figure 13.4

2. Notice that when the graph of the derivative is negative, the function itself is decreasing. When the graph of the derivative is positive, the function itself is increasing. At the point where the graph of the derivative has an x-intercept, the derivative must be zero. This corresponds to the local minimum point of the function where there is a horizontal tangent line.

Exercise 13.3

In Exercises 1-8, find the equation of the tangent line at the given x-value. Graph the function and tangent line on your calculator and make a sketch of the graph and tangent line on your paper. Show values for Xmin, Xmax, Ymin, and Ymax at the ends of the axes of your drawing.

1. $y = x^2 - 7$ at $x = 1.5$
2. $y = 5x - 2x^2$ at $x = 1.1$
3. $y = \dfrac{5}{2 + x^2}$ at $x = 0.75$
4. $y = \dfrac{1}{x^2}$ at $x = -1.25$
5. $y = \sqrt{81 - x^2}$ at $x = 5.00$
6. $y = 25x - 2x^3$ at $x = 2.55$
7. $y = x^3 + 5x^2 - 6x - 4$ at $x = 4.75$
8. $y = 125x - 16x^4$ at $x = 0.532$

In Exercises 9-12, graph the given function and Procedure G26 to determine the derivative at points on the graph.

9. For $y = 8x - x^2$, find the derivative at $x = 1.5$, 2.5, 4.0, 4.5, and 5.5. What is the sign of the derivative for $x < 4.0$? For $x > 4.0$? Is the function increasing or decreasing in each of these regions?
10. For $y = x^3 - 27x$, find the derivative at $x = -5.0$, -4.0, -4.1, -2.5, -1.5, 1.5, 2.0, 2.5, 3.5, and 5.0. What is the sign of the derivative for $x < -3.0$? For $x > -3.0$ and $x < 3.0$? For $x > 3.0$? Is the function increasing or decreasing in each of these regions?
11. For $y = 12x^4 - 6x^2$, find the derivative at $x = -1.0$, -0.6, -0.4, -0.1, 0.1, 0.4, 0.6, and 1.0. For what x-values is the sign of the derivative positive? Negative? For what x-values is the function increasing or decreasing?
12. For $y = \frac{25}{x^2}$, find the derivative at $x = -2.5$, -1.5, -0.5, 0.5, 1.5, and 2.5. For what x-values is the sign of the derivative positive? Negative? For what x-values is the function increasing or decreasing?

In Exercises 13-18, graph the function and its numerical derivative on the same axes as in Example 13.4. In each case, answer these questions: (a) What is true about the function when the graph of the derivative is positive (above the x-axis)? (b) What is true about the function when the graph of the derivative is negative (below the x-axis)? (c) What is true about the function at the x-values that are x-intercepts of the graph of the derivative?

13. $y = 8 - x^2$ What type of graph is obtained for the derivative? Explain.
14. $y = x^3 - 25x^2$ What type of graph is obtained for the derivative? Explain.
15. $y = x^4 - 6x^2$
16. $y = \dfrac{6}{x^2 + 4}$
17. $y = \sqrt{25 - x^2}$ What happens to the graph of the derivative at $x = 5$ and -5? Explain.
18. $y = \dfrac{10}{(x-3)^2}$ What happens to the graph of the derivative at $x = 3$? Explain.

♦ 19. *(Washington, Exercises 24-5, #31)* A batter hits a baseball that follows a path given by $y = x - 0.0025x^2$. Graph the derivative of this function. For what x-value is the derivative zero? What is true at this point? For what x-values is the derivative positive? What is happening to the baseball for these values of x?

♦ 20. *(Washington, Exercises 24-5, #32)* The angle, θ, in degrees, of a robot arm with the horizontal as a function of the time t, in seconds, is given by $\theta = 10 + 12t^2 - 2t^3$ for $0 \le t \le 6$ s. Graph the derivative of this function. For what t-value is the derivative zero? What is true at this point? For what t-values is the derivative negative? What is happening to the angle for these t-values?

In Exercises 21-24, graph the second derivative of the function on the same graph as the function. The second derivative of a function may be graphed by graphing the numerical derivative of the numerical derivative. If Y_1 is the name of the function and $Y_2 = \text{nDeriv}(Y_1, x, x)$, then the second derivative by be represented by $Y_3 = \text{nDeriv}(Y_2, x, x)$. For what values of x is the second derivative negative and for what values of x is the second derivative positive? What is true about the graph of the function for these values? What is true where the second derivative is zero?

21. $y = \frac{x^3}{4}$

22. $y = 2x^2 - 5x$

23. $y = x^4 - 16x^2$

24. $y = 25x - x^3$

Newton's method is a method of finding the roots of an equation $f(x) = 0$ using the formula

$$x_2 = x_1 - \frac{f(x_1)}{f'(x_1)}$$

where x_1 is the first approximation (or guess) of the root, f(x) is the function from the equation, and x_2 is the second approximation. The value x_2 then is used to replace x_1 and a third approximation calculated. The process is repeated until a root of a desired accuracy is obtained. A program that will calculate repeated approximations using Newton's method is:

```
PROGRAM:NEWT
X – Y1 / nDeriv(Y1,X,X) → X
Disp X
```

Before running this program, store the function for the function name Y_1 and the initial guess for X. To obtain successive approximations, keep pressing ENTER before pressing any other keys. (Students, who are so inclined, may modify this program so all this is taken care of in the program.)

In Exercises 25-28, use the program NEWT to find the real roots of the following equations to a precision of 0.000001.

25. $x^2 - 14 = 0$

26. $14 - x^3 = 0$

27. $x^4 - 5x^2 - 8 = 0$

28. $x^3 - 7x^2 + 8 = 0$

Chapter 14

Integrals

14.1 The Indefinite Integral
(Washington, Section 25-3)

A concept just as important as finding the derivative of a function is that of finding the function if we know the derivative. This process is called finding the **antiderivative**. Since the derivative of the function $3x^2$ is 6x, we say that the antiderivative of 6x is $3x^2$. But $3x^2 + 1$, $3x^2 + 2$, $3x^2 - 5$, etc. are also antiderivatives of 6x. In reality there are an infinite number of antiderivatives of 6x, all differing by a constant. Thus, when writing the antiderivative of 6x we write $3x^2 + C$ where C is a constant. (When we take the derivative of $3x^2 + C$, the derivative of C is zero.) In certain cases, enough information may be given to be allow us to determine C.

A notation used to indicate that we are to find the antiderivative of a function f(x) is $\int f(x)\, dx$ and is called an **indefinite integral.** If the derivative of F(x) equals f(x) [F'(x) = f(x)], then the indefinite integral of f(x) equals the function F(x). We write this as

$$\int f(x)\,dx = F(x) + C$$

The function f(x) is called the **integrand** and C is called the **constant of integration**. The process of finding the indefinite integral is called **integration**. Since the antiderivative of 6x is $3x^2 + C$, we write that the integral of 6x equals $3x^2 + C$:

$$\int 6x\,dx = 3x^2 + C.$$

Our interest is to investigate the graph of the antiderivatives of functions and to find the particular graph that satisfies certain given conditions.

<u>Example 14.1</u>: Determine the integral of y = 2x – 3 and graph the resulting function for C = –2, 0, 2, 4, and 6. Determine the equation of the function that passes through the point (3, 2)?

Solution:

1. First determine the indefinite integral (or antiderivative) of $2x - 3$ which gives

$$\int (2x-3)\,dx = x^2 - 3x + C$$

2. Next graph $y = x^2 - 3x + C$ for $C = -2$, $C = 0$, $C = 2$, $C = 4$, and $C = 6$. It may be helpful here to replace C by the list {−2, 0, 2, 4, 6}. (See Procedure C11) The resulting graphs are given in Figure 14.1.
3. The five different graphs lie above each other with the lowest graph given by the smallest value of C. Evaluate the function at $x = 3$ (see Procedure G13) and use the up and down arrow keys to jump from graph to graph until the value 2 is obtained for y. Note which graph this point is on. In this case, the point (3, 2) is on the middle graph where $C = 2$. Thus, the antiderivative that meets the given condition is $y = x^2 - 3x + 2$.

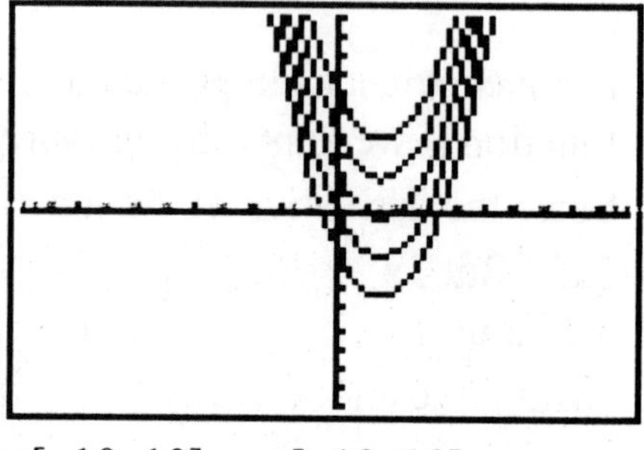

y:[−10, 10] x:[−10, 10]

Figure 14.1

In Example 14.1, it was relative easy to find the right curve. In other cases it may be necessary to zoom in on the graph in the area of the point or to change the viewing rectangle. If a graph doesn't fall exactly on the given point we estimate a value of C and check by graphing the function for this value of C.

A set of curves obtained by graphing a function for different values of a constant contained in the function is called a **family of curves**. In Example 14.1, the indefinite integral of a function gives a family of curves. In many problems, we have certain conditions (such as a point on the graph) that allow us to find the particular curve.

Exercise 14.1

In Exercises 1-10, determine the indefinite integral of the given function and graph the resulting function for the given values of C. Use the graphs to find the equation of that function that passes through the point P.

1. $f(x) = 3x - 7$	$C = -4, -2, 0, 2, 4, 6$	P(2, −4)
2. $f(x) = 5 - 4x$	$C = -4, -2, 0, 2, 4, 6$	P(3, −5)

3. $f(x) = x^2 + 2x$	$C = -5, -3, -1, 1, 3, 5$	P(3, 15)
4. $f(x) = 2x - 3x^2$	$C = -5, -3, -1, 1, 3, 5$	P(2, 1)
5. $f(x) = x^3$	$C = -5, -3, -1, 1, 3, 5$	P(−2, 7)
6. $f(x) = 3x^2 - 4x^3$	$C = -5, -3, -1, 1, 3, 5$	P(1, 5)
7. $f(x) = -2x^{-2}$	$C = -4, -2, 0, 2, 4, 6$	P(−2, 5)
8. $f(x) = 6x^{\frac{1}{2}} + 1$	$C = -4, -2, 0, 2, 4, 6$	P(4, 32)
9. $f(x) = 4(x^2 - 2)^3(2x)$	$C = -4, -2, 0, 2, 4, 6$	P(2, 14)
10. $f(x) = 6(1 - x^3)^5(3x^2)$	$C = -4, -2, 0, 2, 4, 6$	P(−1, 68)

In Exercises 11-14, for each indefinite integral, use graphical methods to find the function that passes through the given point P.

11. $\int (3x^2 + 2x)\,dx$	P(2, 7)
12. $\int (4 - 6x^2)\,dx$	P(1, 5)
13. $\int 8(3x + 5)^7(3)\,dx$	P(−2, 15)
14. $\int (3 - 2x^2)^6(4x)\,dx$	$P(1, \frac{36}{7})$

♦ 15. *(Washington, Exercises 25-3, #39)* The time rate of change of electric current in a circuit is given by $di/dt = 4t - 0.6t^2$. Use graphical means to find the expression for the current as a function of time if i = 2.5 A when t = 1.0 s.

♦ 16. *(Washington, Exercises 25-3, #40)* The time rate of change of velocity (acceleration) of a roller mechanism is $dv/dt = 3\sqrt{t} - 2$. Use graphical means to find the expression for the velocity as a function of time if v = 8 cm/s at t = 4 s.

Exercise 17 is related to the material in Section 14.2 and should be considered and discussed prior to doing the material in Section 14.2.

17. Assume that as a contestant on a TV game show you are placed on an island in the middle of shark infested waters in the Pacific Ocean. The island is of irregular shape and you have in your possession a supply of food and water, a 100 meter tape measure, some wooden stakes, and plenty of string. To get off the island and back to civilization before your food runs out, you must determine the area of the island to within 100 square meters. Explain how you would accomplish this task. Then, explain how you could obtain the area to a higher degree of precision.

14.2 Area Under a Curve
(Washington, Section 25-4)

We next consider the area between the graph of a function $y = f(x)$, the x-axis, and the two lines, $x = a$ and $x = b$. This area is very closely related to integration and is an important concept in the understanding of integrals. One way of approximating such an area is by the use of rectangles as shown in the Figure 14.2. The interval from $x = a$ to $x = b$ along the x-axis is subdivided into smaller intervals called subintervals. If we divide the large interval into N equal subintervals, then each subinterval has a length of $(b - a)/N$ and forms the base of a rectangle. The notation Δx is used to denote the length of each of these subintervals. When using the calculator, we will use H to denote Δx. Thus

$$\frac{b-a}{N} = \Delta x = H$$

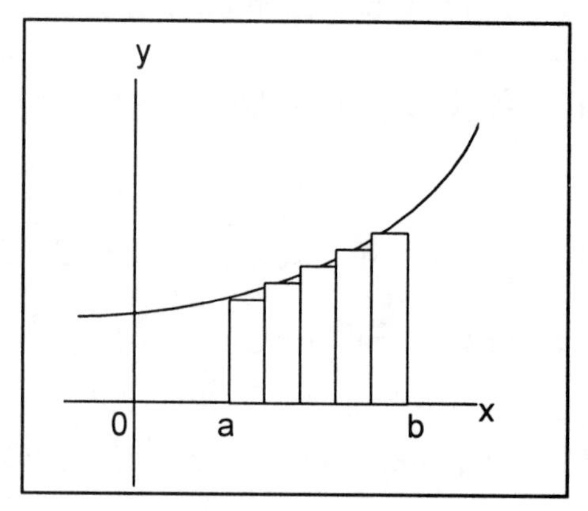

Figure 14.2

The height of each rectangle is determined by selecting an x-value in each subinterval and evaluating the function f(x) at this value. The height depends on the value of x and on the function used to calculate the height. The area of each rectangle is found by multiplying the base of each rectangle by its height. The sum of the areas of all the rectangles with the subintervals as bases is an approximation to the total area. For convenience, we will determine the heights of the rectangles by evaluating the function at either the left end or the right end of each subinterval.

Example 14.2: Use 3 rectangles to estimate the area under the graph of $y = x^2 + 2$, above the x-axis, and between the lines $x = 1$ and $x = 4$. Do calculations (by hand) by first using the left end of each subinterval and then by using the right hand of each subinterval.

Solution:

1. The graph of $y = x^2 + 2$, with the left end of each subinterval used to determine the height of the rectangle, is shown in Figure 14.3. To do these calculations we divide the interval from $x = 1$ to $x = 4$ into 3 equal subintervals, each of one unit length

$$\frac{4-1}{3} = 1 = H$$

2. The left endpoints of each subinterval are 1, 2, and 3. Evaluating the function at the left endpoints gives f(1) = 3, f(2) = 6, and f(3) = 11 for the heights of the rectangles. The area of each rectangle is found by multiplying its width 1 by its height. The area of the first rectangle is $3\times1 = 3$, the next is $6\times1 = 6$, and the third is $11\times1 = 11$. The sum is

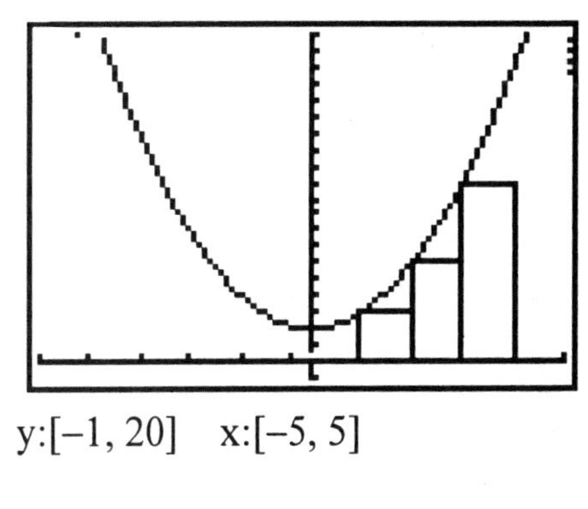

y:[−1, 20] x:[−5, 5]

Figure 14.3

$$\begin{aligned} 3\times1 + 6\times1 + 11\times1 &= 3 + 6 + 11 \\ &= 20 \text{ square units.} \end{aligned}$$

3. For the right end points we evaluate the function at 2, 3, and 4. This gives f(2) = 6, f(3) = 11, and f(4) = 18. The area of the first rectangle is $6\times1 = 6$, the second $11\times1 = 11$, and the third $18\times1 = 18$. The sum is

$$\begin{aligned} 6\times1 + 11\times1 + 18\times1 &= 6 + 11 + 18 \\ &= 35 \text{ square units.} \end{aligned}$$

For this function the rectangles using the left hand endpoints all lie under the graph and give a lower bound for the areas -- an value that is less than the exact area. Thus, the exact area of this region must be greater than 20 square units. With the right hand endpoints, the area under the curve lies within the total area of all the rectangles and the calculated area must be an upper bound -- an area that is greater than the exact area. The exact area of the region is less than 35 square units.

In Example 14.2, the lower and upper bound on the area and thus, the estimate of the area, may be improved by taking a larger number of rectangles (a smaller base of each rectangle). However, this would require many more calculations. Such calculations are better performed using a calculator or computer.

We now consider a program for the graphing calculator that will both draw the rectangles and find the area of the rectangles. First we look at how to access some additional programming commands. It is not the intention to give an in-depth discussion of programming, but only to provide enough understanding to enter and use the given program.

Procedure P8. To display a graph from within a program.

While in programming edit mode (at a step in program):

On the TI-82:
1. Press the key PRGM
2. Select I/O
3. Select 4:DispGraph

On the TI-85:
1. Select I/O
2. Select DispG

A loop in a program may be constructed on most graphing calculators by using the label, if, and goto statements. A Goto statement causes the execution of a program to transfer to another statement identified by a label statement. The If statement determines when a program is to transfer by determining if a certain *condition* is met. A condition often involves the mathematical symbols <, ≤, >, ≥, = or ≠ (Procedure C13). For example, consider the following If statement in a program: If k < 5. As long as k is less than 5 when execution of the program reaches this statement, the next statement in the program will be executed. However, if k is not less than 5, the next statement will be skipped and the following statement executed.

Procedure P9. Creating a loop (Lbl, If, and Goto commands).

While in programming edit mode (at a step in program):

On the TI-82:
1. Select the label (Lbl) command:
   a. Press PRGM
   b. With CTL highlighted, select 9:Lbl
2. Select the If command:
   a. Press PRGM
   b. With CTL highlighted, select 1:If
3. Select the Goto command:
   a. Press PRGM
   b. With CTL highlighted, select 0:Goto

Note: A label may be a letter or a numerical digit.

On the TI-85:
1. Select the label (Lbl) command:
   a. Select CTL
   b. Select Lbl (after pressing MORE)
2. Select If command
   a. Select CTL
   b. Select If
3. Select the Goto command
   a. Select CTL
   b. Select Goto (after pressing MORE)

Note: A label may be a name of up to 8 characters, beginning with a letter.

The form of the label command is Lbl *label*
The form of the If command is If *condition*
The form of the Goto command is Goto *label*

The construction of the loop is of the form:

... (program statements before loop)
Lbl *label*
... (program statements within loop)
...
If *condition*
Goto *label*
... (programming statements after loop)

Procedure P10. Drawing a line on the screen from within a program (Line command).

While in programming edit mode (at a step in program):

On the TI-82:
1. Press the key DRAW
2. With DRAW highlighted, select 2:LINE

On the TI-85:
1. Press the key GRAPH
2. Select DRAW (after pressing MORE)
3. Select LINE

The form of the line command is LINE(*a, b, c, d*) where a and b are the coordinates of one point and c and d are the coordinates of a second point. This command draws a line from (a, b) to (c, d).
Example: LINE $(0,2,X,Y_1)$ draws a line from the point (0,2) to the point (x,f(x)).

Procedure P11. To temporarily halt execution of a program. (PAUSE statement)

While in programming edit mode (at a step in program):

On the TI-82:
1. Press the key PRGM
2. With CTL highlighted, select 8:PAUSE

On the TI-85:
1. Select CTL
2. Select Pause (after pressing MORE twice)

The PAUSE statement when used in a program halts execution of the steps of a program until the ENTER key is pressed.

The following program asks for values of A, B, and N to be entered and then draws N rectangles using the left end points of each subinterval. After the graph is drawn, press ENTER to see the value for the area of the N rectangles. The arrow (→) refers to the STO▷ key. Before executing the program the function, must be stored for Y_1.

```
PROGRAM: ARECTG1
ClrDraw
Prompt A
Prompt B
Prompt N
0 → S
(B-A)/N → H
DispGraph
A → X
Lbl A
LINE(X, 0, X, Y1)
LINE(X + H, 0, X + H, Y1)
LINE(X, Y1, X + H, Y1)
S + Y1*H → S
X + H → X
If X < B
Goto A
DispGraph
PAUSE
Disp "AREA IS"
Disp S
End
```

Example 14.3: Use the program ARECTG1 to estimate the area under the graph of $y = x^2 + 2$, above the x-axis, and between $x = 1$ and $x = 4$ using left end points of the subintervals. Do the calculation using 3, 6, and 12 rectangles.

Solution:

1. Make sure the function $x^2 + 2$ is entered for Y_1 and all other functions are deleted or turned off and be sure the program ARECTG1 is correctly entered into your calculator. Execute the program (See Procedure P2). After the graph is drawn, press ENTER to see the calculation for the area.
2. Enter 1 for A, 4 for B, and 3 for N. The resulting graph is the same as in Figure 14.3 and the area is 20 square units.
3. Execute the program a second time using A = 1, B = 4, and N = 6. This time the area of the rectangles is 23.375 square units.
4. Execute the program a third time using A = 1, B = 4, and N = 12. This time the area of the rectangles is 25.15625 square units. (The exact area is 27 square units.)

Exercise 14.2

In Exercises 1-8, use the program ARECTG1 to graph the given function and determine the specified areas for the given values of N.

1. The area bounded by the function $y = 2x + 2$, the x-axis, $x = 1$ and $x = 5$. Determine the area of this region by using the calculator with $N = 2, 4$, and 8, and also by hand by finding the area of the resulting trapezoid.
2. The area bounded by the function $y = 5 - x$, the x-axis, $x = -2$, and $x = 4$. Determine the area of this region by using the calculator with $N = 2, 4$, and 8 and also by hand by finding the area of the resulting trapezoid.
3. The area bounded by the function $y = 9 - x^2$, the x-axis, $x = 0$, and $x = 3$ for $N = 2, 4, 8$, and 16. Are the values obtained upper bounds or lower bounds for the exact area? Make a guess as to the exact area (you may wish to use larger values for N).
4. The area bounded by the function $y = x^2 + 5$, the x-axis, $x = -3$, and $x = 3$ for $N = 2, 4, 8$, and 16. Are the values obtained upper bounds or lower bounds for the exact area? Make a guess as to the exact area (you may wish to use larger values for N).
5. The area bounded by the function $y = x^3$, the x-axis, $x = 0$, and $x = 3$ for $N = 2, 4$, 8, and 16. Are the values obtained upper bounds or lower bounds for the exact area? Make a guess as to the exact area (you may wish to use larger values for N).
6. The area bounded by the function $y = 27 - x^3$, the x-axis, $x = -1$, and $x = 3$ for $N = 2, 4, 8$, and 16. Are the values obtained upper bounds or lower bounds for the exact area? Make a guess as to the exact area (you may wish to use larger values for N).
7. The area bounded by the function $y = \frac{1}{x^2}$, the x-axis, $x = 1$, and $x = 8$ for $N = 2, 4$, 8, and 16. Are the values obtained upper bounds or lower bounds for the exact area? Make a guess as to the exact area (you may wish to use larger values for N).
8. The area bounded by the function $y = \sqrt{x+2}$, the x-axis, $x = -2$, and $x = 2$ for $N = 2, 4, 8$, and 16. Are the values obtained upper bounds or lower bounds for the exact area? Make a guess as to the exact area (you may wish to use larger values for N).
9. Modify the program ARECTG1 so that right end points of each subinterval will be used to calculate the area of the rectangles. Call the resulting program ARECTG2.
10. Modify the program ARECTG1 so that the mid-point of each subinterval is used to calculate the area of the rectangles. Call the resulting program ARECTGM.
11. Write a program ARECT1 that will calculate the area using rectangles and left hand end points, but which does not show the rectangles or graphs.

12. Write a program to calculate the area of rectangles using the left end points of each subinterval and the right end points of each subinterval and print out both values. This program should not show the rectangles or graphs.
13. Use the program ARECTG2 to do Exercise 1. Using the results and the results of problem 1, give upper and lower bounds for the exact area.
14. Use the program ARECTGM to do Exercise 2.
15. Use the program ARECTG2 to do Exercise 3. Using the results and the results of problem 3, give upper and lower bounds for the exact area.
16. Use the program ARECTGM to do Exercise 4.
17. Use the program ARECTG2 to do Exercise 5. Using the results and the results of problem 5, give upper and lower bounds for the exact area.
18. Use the program ARECTGM to do Exercise 6.
19. Use the program ARECTG2 to do Exercise 7. Using the results and the results of problem 7, give upper and lower bounds for the exact area.
20. Use the program ARECTGM to do Exercise 8.
21. Use the program ARECTG1 to estimate the area bounded by $y = 3x^2$, the x-axis, $x = 1$ and $x = 3$. Use N values of 5, 10, 50, and 100.
    (a) Are the areas for these N-values approaching a limit? If so, what do you think it is?
    (b) Find the antiderivative of $3x^2$. Evaluate this antiderivative at $x = 1$ and subtract from the value of the antiderivative at $x = 3$. How does this number compare to your answer in part (a)?
22. Use the program ARECTG1 to estimate the area bounded by $y = 4 - x^2$, the x-axis, $x = -2$ and $x = 2$. Use N values of 5, 10, 50, and 100.
    (a) Are the areas for these N-values approaching a limit? If so, what do you think it is?
    (b) Find the antiderivative of $4 - x^2$. Evaluate this antiderivative at $x = -2$ and subtract from the value of the antiderivative at $x = 2$. How does this number compare to your answer in part (a)?
23. Use the program ARECTG1 to estimate the area bounded by $y = x^3$, the x-axis, $x = 0$ and $x = 4$. Use N values of 5, 10, 50, and 100.
    (a) Are the areas for these N-values approaching a limit? If so, what do you think it is?
    (b) Find the antiderivative of x^3. Evaluate this antiderivative at $x = 0$ and subtract from the value of the antiderivative at $x = 4$. How does this number compare to your answer in part (a)?
24. Use the program ARECTG1 to estimate the area bounded by $y = x^2 + 5$, the x-axis, $x = -1$ and $x = 2$. Use N values of 5, 10, 50, and 100.
    (a) Are the areas for these N-values approaching a limit? If so, what do you think it is?

(b) Find the antiderivative of $x^2 + 5$. Evaluate this antiderivative at $x = -1$ and subtract from the value of the antiderivative at $x = 2$. How does this number compare to your answer in part (a)?

25. Use the program ARECTG1 to estimate the area bounded by $y = x^2 - 16$, the x-axis, $x = 0$ and $x = 3$ for $N = 2, 4,$ and 8. Does the calculated area of the rectangles come out positive or negative? Can you explain why?
26. Use the program ARECTG1 to estimate the area bounded by $y = x^2 - 4x - 5$, the x-axis, $x = 0$ and $x = 4$ for $N = 2, 4,$ and 8. Does the calculated area of the rectangles come out positive or negative? Can you explain why?
27. (a) Use the program ARECTG1 to estimate the area bounded by $y = 8x - x^2$, the x-axis, $x = 0$ and $x = 8$ for $N = 10$ and 50.
    (b) Repeat for the interval $x = 0$ to $x = 12$. Explain the results obtained.
    (c) Repeat for the interval $x = 0$ to $x = 8$ and for the interval $x = 8$ to $x = 12$. How do these two values relate to the answer in part (b)?
28. (a) Use the program ARECTG1 to estimate the area bounded by $y = x^2 - 5x - 6$, the x-axis, $x = 0$ and $x = 6$ for $N = 10$ and 50.
    (b) Repeat for the interval $x = 0$ to $x = 10$. Explain the results obtained.
    (c) Repeat for the interval $x = 0$ to $x = 6$ and for the interval $x = 6$ to $x = 10$. How do these two values relate to the answer in part (b)?
29. Use the program ARECTG1 to estimate the area bounded by $y = \sin x$, the x-axis, $x = 0$ and $x = \frac{\pi}{2}$ for $N = 5, 10, 50$ and 100. (Calculator should be in radian mode.)
30. (a) Use the program ARECTG1 to estimate the area bounded by $y = \frac{1}{x}$, the x-axis, $x = 1$ and $x = 2$ for $N = 5, 10, 50,$ and 100. Then find ln 2 on your calculator using the LN key. How to the results compare?
    (b) Repeat part (a) for the interval $x = 1$ to $x = 3$ and compare the results to ln 3.
    (c) What can you conclude about the area under the graph of $y = \frac{1}{x}$ from $x = 1$ to $x = K$ and ln K?

14.3 The Definite Integral and Trapezoidal and Simpson's Rules *(Washington, Sections 25-5, 25-6, and 25-7)*

From Section 14.2, we conclude that we can estimate the area under a curve, above the x-axis and between two values of x by considering the areas of a large number of rectangles in this region. The precision to which we can calculate the area is limited only by the number of rectangles selected and the corresponding calculations required. If we could take an infinite number of rectangles, we would obtain such an area exactly. This area is also related to the antiderivative (or integral) of the given function (see Exercises 21-24 of Exercise Set 14.2).

We sum this up in the following statement: If F(x) is the antiderivative of f(x), then $\int_a^b f(x)\,dx$ represents the area of the region bounded by the graph of y = f(x), the x-axis, x = a and x = b. An integral of the form $\int_a^b f(x)\,dx$ is called a **definite integral.** The value a is referred to as the **lower limit** of integration and b is referred to as the **upper limit** of integration.

This definite integral may be evaluated by finding the antiderivative of f(x) and subtracting the value of the antiderivative at x = a from the value of the antiderivative at x = b. This last statement is summarized as

$$\int_a^b f(x)\,dx = F(x)\Big|_a^b = F(b) - F(a) \qquad \text{where } F'(x) = f(x) \quad \text{[Equation 1]}$$

The notation $F(x)\Big|_a^b$ means first evaluate F(x) at x = b, then subtract the value at x = a.

If we can find antiderivative of f(x) exactly then we can determine the value of the definite integral exactly. However, it is often difficult or impossible to determine the antiderivative of a function directly. Knowing that the definite integral represents an area allows us to approximate the value of a definite integral by numerical means. We have already used one such method -- that of estimating areas by using rectangles. The method of using rectangles is not good because it takes a large number of calculations to achieve a high precision. In this section we consider two other methods of approximating this area.

Even though we are considering the definite integral as representing the area under a graph, applications involving the definite integral often have no direct relationship to an area. For example, definite integrals may be used to find volumes, amount of work or force, and centers of gravity.

One method of estimating the value of a definite integral is to think of it as a area and divide the interval from x = a to x = b up into equal subintervals each of length Δx as we did with rectangles. This time we use trapezoids and come up with a formula for estimating the area. We will not develop this formula here, but only state it as the **trapezoidal rule**:

$$A_T = \frac{H}{2}(y_0 + 2y_1 + 2y_2 + \ldots + 2y_{n-1} + y_n) \qquad \text{[Equation 2]}$$

where $H = \Delta x =$ the length of a subinterval $= \dfrac{b-a}{2}$ and $y_0 = f(a)$, $y_1 = f(a+H)$, $y_2 = f(a+2H)$, etc. The last y-value $y_n = f(b)$. The y-values represent the distance from the x-axis to the graph of the function at the ends of the each subinterval.

The following program estimates the value of a definite integral using the trapezoidal rule.

```
PROGRAM: TRAPRULE
0 → S
1 → K
Prompt A
Prompt B
Prompt N
(B–A)/N → H
A → X
Y1 → S
Lbl A
A + K*H → X
S + 2Y1 → S
K + 1 → K
If K ≤ (N–1)
Goto A
B → X
Y1 + S → S
S*H/2 → S
Disp "Value is"
Disp S
```

If the exact value of the definite integral is known, we may compare it with the answer found using the trapezoidal rule and may calculate the percentage error. The percentage error is found using the formula

$$\text{Percent Error} = \left|\frac{\text{Exact value} - \text{Calculated value}}{\text{Exact value}} \times 100\right| \qquad \text{[Equation 4]}$$

<u>Example 14.4</u>: Use the program TRAPRULE to estimate $\int_1^4 (x^2+2)\,dx$. Use N = 6, 12, and 100. Compare answers to those in Example 14.4. Determine the percentage error in each case.

Solution:

1. Make sure that the function x^2+2 is entered for Y_1 and all other functions are deleted or turned off and be sure the program TRAPRULE is correctly entered into your calculator. Execute the program.
2. Enter 1 for A, 4 for B, and 6 for N. The result is 27.125. The value obtained in Exercise 14.4 for N = 6 was 23.375. This value is much closer to the exact value of 27.0 and gives a percentage error of

$$\left|\frac{27.0-27.125}{27.0}\times 100\right| = 0.463\%.$$

3. Enter 1 for A, 4 for B, and 12 for N. This time the result is 27.03125 which gives a percentage error of 0.116%. In Exercise 14.4 the value for N = 12 was 25.25625.
4. Enter 1 for A, 4 for B, and 100 for N. The result is 27.00045 which is within 0.001 units of the exact value for a percentage error of 0.0017%.

In Example 14.4, we see that the trapezoidal rule is much more precise than using rectangles. A much smaller value of N needs to be taken to obtain a reasonable degree of precision.

An even more precise method of estimating a definite integral is the use of Simpson's rule. Simpson's rule approximates the area under a curve by approximating pieces of the curve using parabolas and considering the area under the parabolas. Again we will not develop the formula here, but just state **Simpson's rule**:

$$A_s = \frac{H}{3}(y_0+4y_1+2y_2+4y_3+2y_4+...+4y_{n-1}+y_n) \qquad \text{[Equation 5]}$$

Again $H = \Delta x = \dfrac{b-a}{N}$ and $y_0 = f(a)$, $y_1 = f(a+H)$, $y_2 = f(a+2H)$, etc. When using Simpson's rule, N must be an even number. The program for Simpson's rule is left as an exercise.

Exercise 14.3

In Exercises 1-8, use the program TRAPRULE to estimate the definite integrals. Compare your answers to those found to the related problem in Exercise 14.2. Use Equation 1 and the antiderivative to find the exact value and then use Equation 4 to find the percentage error in your values.

1. $\int_1^5 (2x+2)\,dx$ for $N = 4$ and 8

2. $\int_{-2}^4 (5-x)\,dx$ for $N = 4$ and 8

3. $\int_0^3 (9-x^2)\,dx$ for $N = 4$ and 16

4. $\int_{-3}^3 (x^2+5)\,dx$ for $N = 4$ and 16

5. $\int_0^3 x^3\,dx$ for $N = 4$ and 16

6. $\int_{-1}^3 (27-x^3)\,dx$ for $N = 4$ and 16

7. $\int_1^8 \frac{1}{x^2}\,dx$ for $N = 8$ and 16 (Exact value is 0.875)

8. $\int_{-2}^2 \sqrt{x+2}\,dx$ for $N = 8$ and 16 (Exact value is $\frac{16}{3}$.)

9. Evaluate $\int_1^2 (4x^3+2)\,dx$ using the program TRAPRULE for $N = 2, 4, 8$, and 16. Then determine the exact value and the percentage error in each case. When we double the value of N by what factor does this decrease the percentage error? Does this agree with your results in Exercises 1-8?

10. Evaluate $\int_0^1 (4x-x^4)\,dx$ using the program TRAPRULE for $N = 2, 4, 8$, and 16. Then determine the exact value and the percentage error in each case. When we double the value of N by what factor does this decrease the percentage error? Does this agree with your results in Exercises 1-8?

11. Use the program TRAPRULE to find each of the integrals $\int_{-2}^2 (x^3-4x)\,dx$, $\int_{-2}^0 (x^3-4x)\,dx$, and $\int_0^2 (x^3-4x)\,dx$ for $N = 20$. Explain the results. Write an equation involving these three integrals.

12. Use the program TRAPRULE to find each of the integrals $\int_{-3}^{3}(9x-x^3)\,dx$, $\int_{-3}^{0}(9x-x^3)\,dx$, and $\int_{0}^{3}(9x-x^3)\,dx$ for $N = 20$. Explain the results. Write an equality involving these three integrals.

13. Use the program TRAPRULE to find each of the integrals $\int_{1}^{3}6x^3\,dx$ where $A = 1$ and $B = 3$ and $\int_{3}^{1}6x^3\,dx$ where $A = 3$ and $B = 1$. In both cases use $N = 20$. What is the relationship of these two integrals. Write an equality involving these two integrals.

14. Use the program TRAPRULE to find each of the integrals $\int_{1}^{4}\frac{1}{x^2}\,dx$ where $A = 1$ and $B = 4$ and $\int_{4}^{1}\frac{1}{x^2}\,dx$ where $A = 4$ and $B = 1$. In both cases use $N = 20$. What is the relationship of these two integrals. Write an equality involving these two integrals.

15. Write a program called SIMPRULE for calculating the definite integral using Simpson's rule. It should be similar to the program TRAPRULE but take into account the differences in the formula.

16. Use the program SIMPRULE to evaluate $\int_{1}^{4}\frac{1}{x}\,dx$ for $N = 10$. Compare the results to the exact answer ln 4. What is the percentage error?

17.-24. Do each of the Exercises 1-8 using the program SIMPRULE instead of TRAPRULE.

Do Exercises 25-28 using both TRAPRULE and SIMPRULE programs.

♦ 25. *(Washington, Exercises 25-5, #33)* The work W (in ft-lb) in winding up an 85-ft cable is given by $W = \int_{0}^{85}(1000-5x)\,dx$. Find W. Use $N = 10$.

♦ 26. *(Washington, Exercises 25-5, #36)* The total force (in Newtons) on the circular end of a water tank is given by $F = 19600\int_{0}^{6}y\sqrt{36-y^2}\,dy$. Find F. Use $N = 10$.

♦ 27. *(Washington, Exercises 25-6, #15)* A force F that a distributed electric charge has on a point charge is $F = k\int_0^2 \frac{dx}{(4+x^2)^{\frac{3}{2}}}$ where x is the distance along the distributed charge and k is a constant. Find F in terms of k. Use N = 10.

♦ 28. *(Washington, Exercises 25-7, #16)* The average value of the electric current i_{av}, in amperes in a circuit for the first 5.5 s is given by $i_{av} = \frac{1}{5.5}\int_0^{5.5} (4t - t^2)^{0.2}\, dt$. Find i_{av} using N = 10.

14.4 More on Integration and Areas *(Washington, Section 26-2)*

Many graphing calculators have a built in function that numerically calculates the definite integral. In this section, we make use of this built in function along with some other features of the calculator to determine areas.

Procedure G27. Function to determine the area under a graph.

With the graph of a function on the screen:

On the TI-82:
1. Press the key CALC
2. Select 7:$\int f(x)dx$
3. Move the cursor to the first x-value and press ENTER.
4. Move the cursor to the second x-value and press ENTER.

(The area will be shaded.)

On the TI-85:
1. Select MATH (after pressing MORE)
2. Select $\int f(x)$
3. Move the cursor to first x-value and press ENTER.
4. Move the cursor to the second x-value and press ENTER.

The area between the curve, the x-axis, and the two x-values will be given at the bottom of the screen.

Procedure C20. Function to calculate definite integral.

On the TI-82:
1. Press the key MATH
2. Select 9:fnInt(

On the TI-85:
1. Press the key CALC
2. Select fnint

The form of the function if fnInt(y, variable, a, b) where y is a function or name of a function, variable is the variable used in the function, and a and b are the lower and upper limits of integration.

(Example: fnInt(x^3, x, 1,3) is used to evaluate $\int_1^3 x^3\,dx$.)

Example 14.5: Evaluate $\int_1^4 (x^2+2)\,dx$.

Solution:

1. Enter x^2+2 for Y_1 and graph this function. In this case, it better to use the decimal viewing rectangle but with Ymax = 20.
2. Use Procedure G27 to determine the area between x = 1 and x = 4.

   or

   Use Procedure C20 to determine the definite integral between x = 1 and x = 4.

With either procedure, we obtain the exact answer of 27.0. Had we not used the decimal viewing rectangle, it would have been difficult to enter the x-values exactly in Procedure G27.

To find the area between two curves, we not only need to know the two curves, but to also know the points of intersection of the curves. We should apply one of the following to determine the area between two curves.

Between x = a and x = b with interval along x-axis:

$$\int_a^b (\text{top function of } x - \text{bottom function of } x)\,dx$$

Between y = c and y = d with interval along y-axis:

$$\int_c^d (\text{right function of } y - \text{left function of } y)\,dy$$

Graphing procedures could not be used on this last form unless we change the variable y to x and think of the integrand as a function of x. Example 14.6 will demonstrate a method to find the area between two curves where the area is between points of intersection.

Example 14.6: Find the area between the graphs of $y = x^2$ and $y = 4 - x^2$.

Solution:

1. Enter the function x^2 for Y_1 and $4 - x^2$ for Y_2. Then graph so that the whole area appears on the screen. On the standard viewing rectangle, the graph is as in Figure 14.4.

2. Use Procedure G15 to determine the left most point of intersection and enter X→A to store the x-value for the variable A.
3. Use Procedure G15 to determine the right most point of intersection and enter X→B to store this x-value for the variable B.

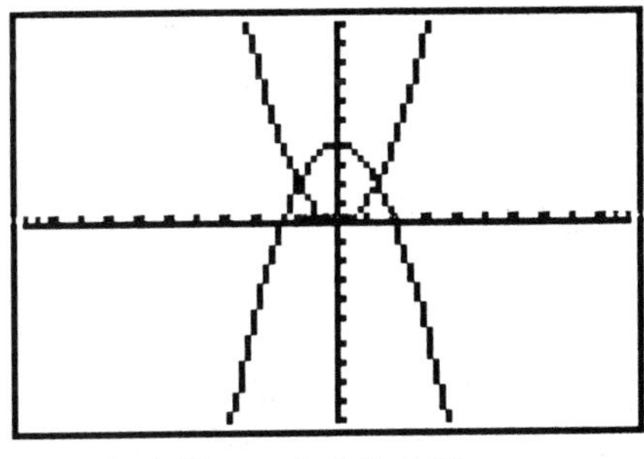

y:[−10, 10] x:[−10, 10]

Figure 14.4

4. Use Procedure C20 to find the definite integral between the two curves and between x = A and x = B.
   Use fnInt(Y_2 − Y_1, X, A, B). This gives the area as 7.5425 rounded to five significant digits.

Exercise 14.4

In Exercises 1-8, use the built in function to determine the following definite integrals.

1. $\int_{1.5}^{2.7} x^2\,dx$
2. $\int_{-1.8}^{1.8} (x^3 - 2)\,dx$
3. $\int_{-1.0}^{1.5} (x - \frac{1}{x})\,dx$
4. $\int_{-0.25}^{1.75} (x^4 - 3x^2 + 2)\,dx$
5. $\int_{1.15}^{1.38} 2^{x^3}\,dx$
6. $\int_{-0.5}^{0.5} \frac{1}{1+x^2}\,dx$
7. $\int_{-5}^{5} \sqrt{25 - x^2}\,dx$
8. $\int_{-\sqrt{2}}^{\sqrt{2}} (x^3 + 5x^2 - 2x - 10)\,dx$

In Exercises 9-20, use the methods of this section to find the areas bounded by the indicated curves.

9. $y = 3x^2$, $y = 0$, and $x = 2.5$
10. $y = 4 - x^2$, $x = 0$, and $y = 0$
11. $y = x^2 - 4$ and $y = x + 2$
12. $y = 2x^2 + 1$ and $y = x + 7$
13. $y = x^2$, $y = 4$, and $x = 0$ $(x > 0)$
14. $y = x^3$, $y = 27$, and $x = 0$
15. $y^2 - 5y = x$ and $y = 2x$
16. $y = 3x - x^2$, $x = 0$ and $y = 1.8$
17. $y = 2\sqrt{x-2}$, $y = 0.5x$, and $y = 0$
18. $y = x^4$, $y = 5 - x$, and $y = 0$
19. $y = x^3$ and $y = 27 - 27x + 9x^2 - x^3$

20. $y = x^2 + 20$, $y = x^2 - 10x + 45$, and $y = 4x^2 - 20x + 40$ for $y < 26.25$
21. The end of a building consists of two parallel walls 20.0 feet high at $x = 0.0$ and $x = 40.0$ with a roof in the shape of a parabola given by $y = 20 + x - 0.025x^2$. Find the area of the end of the building.
22. The boundary of a piece of land on the curve of a road when sketched on a coordinate system has as its boundaries the circle of radius 6 meters, the line $3x + 4y = 120$ and the x- and y-axes where x and y are measured in meters. Find the area of this piece of land.

♦ 23. *(Washington, Exercises 26-2, #31)* Since the vertical displacement s, velocity v, and time t, of a moving object are related by $s = \int v\,dt$, it is possible to represent the change in displacement as an area. A rocket is launched such that its vertical velocity v (in km/s) as a function of time in seconds is $v = 1.25 - 0.015\sqrt{2t+1}$. Find the change in vertical displacement from $t = 9.0$ s to 98.0 s.

♦ 24. *(Washington, Exercises 26-2, #34)* Using CAD (computer-assisted design), an architect programs a computer to sketch the shape of a swimming pool designed between the curves $y = \frac{825x}{(x^2+10)^2}$, $y = -\frac{825x}{(x^2+10)^2}$, and $x = 6.5$. All dimensions are in meters. Find the area of the surface of the pool. If the pool is uniformly two meters deep, how many cubic meters of water are needed to fill the pool?

Appendix A

Procedures for TI-81, TI-82 and TI-85 Calculators

Procedures are given for the TI-82 and TI-85 graphing calculators. For the TI-81 calculator, generally use the procedures for the TI-82. Where differences occur, they are indicated by the symbol ❖.

A.1 Basic Calculator Procedures

Procedure B1. To turn calculator on and off.

1. To turn the calculator on, press the key: ON (in the lower left corner of the calculator).
2. To turn the calculator off, you will need to first press the key: 2nd then the key: ON

Procedure B2. To clear all memory in the calculator.

WARNING: The following steps will erase ALL calculator memory.

1. Turn the calculator on.
2. Press the key: MEM (Do by pressing the key: 2nd then the key: +)

On the TI-82:
A menu will appear on the screen indicating

1:Check RAM...
2:Delete...
3:Reset...

Press the key 3 to reset.
Press the key 2 to erase everything and key 1 ti return to calculations screen.

On the TI-85:
Press the key F3 for RESET.
Press F1 for ALL.
Press F4 to erase everything and F5 to return to calculations screen.

Note: To check the amount of memory available:

On the TI-82:

From:

1:Check RAM...
2:Delete...
3:Reset...

Press the key 1 to check memory.
The amount of memory available is given in terms of units of memory called Bytes. The TI-82 has about of 29000 bytes of memory available.

❖ For TI-81, press the key RESET and select 2:Reset.

On the TI-85:

Press the key F1 for RAM.

Procedure B3. Adjusting the contrast of the screen.

To adjust the contrast of the screen you will need to use the key: 2nd and the arrow keys which appear on the upper right of the keyboard.

1. To make the screen lighter:
   Press the key 2nd, then press the bottom arrow key.
2. To make the screen darker:
   Press the key 2nd, then press the top arrow key.

The contrast change takes place in a series of increments. Thus, the above steps may have to be repeated to obtain the contrast you desire. Note that the key: 2nd must be pressed each time before pressing the arrow key.

Procedure B4. To Clear the Calculator Screen

1. To clear the calculator screen, press the key: CLEAR twice.
2. In some cases, pressing the CLEAR key will not clear the screen. To clear the screen, it is necessary to exit the given screen. Do this by pressing the key: QUIT

A.2 General Calculator Procedures

Procedure C1. To perform secondary operations or obtain alpha characters.

1. To perform secondary operations:
   a) Press the key: 2nd
   b) Then, press the key for the desired action
   (For example: To find the square root of a number, first press the key: 2nd followed by the key: x^2. Then, enter the number and press ENTER.)
2. To obtain alpha characters:
   a) Press the key: ALPHA
   b) Then, press the key for the desired character.

(For example: To display the letter A on the screen, press the key: ALPHA then press the key MATH or the TI-82 or the key LOG on the TI-85)

Note: The alpha key may be locked down on the TI-82 by first pressing the key 2nd then pressing the key: ALPHA. On the TI-85, press the key ALPHA twice.

Procedure C2. **Selection of menu items**

A menu on the TI-82 calculator appears as a list of numbers, each followed by a colon and a menu item. A menu item may be selected by:

1. Pressing the number of the item desired

or

2. Moving the highlight to the item desired by pressing the up and down cursor keys, then pressing the key ENTER.

(Example: To find the cube root of 3.375, press the key MATH, leave MATH highlighted in the top row and select 4:$\sqrt[3]{}$. Then enter the number 3.375 and press ENTER. The result is 1.5.)

In some cases more than one menu may be accessed by pressing the same key. This is indicated by more than one title appearing at the top of the screen with one of the words highlighted. As the right and left arrow keys are used to move the highlight from one title to another. different menus appear on the screen. Thus, first select the correct menu, then the desired item.

A menu on the TI-85 calculator appears as a list on the bottom of the screen. A selection is made by pressing the function key just below the item desired. The function keys are the keys labeled F1, F2, F3, F4, and F5.

In some cases, election of a menu item will produce a secondary menu. Items are selected from the secondary menu by pressing the appropriate function key.

If the symbol → appears on the right of a menu, there are additional items in the menu. To see these additional items, press the key: MORE

To exit from a menu to the previous menu or to the calculations screen, press the key: EXIT

(Example: To find the largest integer less than or equal to 2.095, press the key: MATH, then press the key F1 for NUM, and from the secondary menu, select **int** by pressing the key F4. Then enter the number 2.095 and press ENTER. The result is 2.)

(Example: To find the largest integer which is less than or equal to the number 2.15, first press the key MATH, move the highlight in the top row to NUM, select 4:Int, enter the number 2.15, and press ENTER. The answer is 2.)

To exit from a menu without selecting an option, press QUIT.

Procedure C3. To perform arithmetic operations.

1. To add, subtract, multiply or divide press the appropriate key: $+, -, \times$, or $\div$. On the graphing calculator screen, the multiplication symbol will appear as * and the division symbol will appear as /. If parenthesis are used to denote multiplication it is not necessary to use the multiplication symbol. (The calculation 2(3 + 5) is the same as 2×(3 + 5).)
2. To find exponents use the key ^
   (Example: 2^4 is entered as 2^4)
   For the special case of square we may use the key x^2.
3. After entering the expression press the key ENTER to perform the calculation.

Notes: a. On the graphing calculator, there is a difference between a negative number and the operation of subtraction. The key (-) is used for negative numbers and the key – is used for subtraction.

b. Enter numbers in scientific notation by using the key EE. (Example: To enter $2.5\times\ 0^5$, enter 2.15, press EE, enter 5.)

Procedure C4. To correct an error or change a character within an expression.

1. (a) To make changes in the current expression: Use the left cursor key to move the cursor to the position of the character to be changed.
   (b) To make changes in the last expression entered, after the ENTER key has been pressed, obtain a copy of the expression by pressing the key ENTRY. Then, use the cursor keys to move the cursor to the character to be changed.
2. Characters may now be replaced, inserted, or deleted at the position of the cursor.
   (a) To replace a character at the cursor position, just press the new character.
   (b) To insert a character or characters in the position the cursor occupies,
   Press the key: INS.
   Then, press keys for the desired character(s).

(c) To delete a character in the position the cursor occupies, press the key: DEL

Procedure C5. Selection of special mathematical functions.

1. To raise a number to a power:
   Enter the number
   Press the key ^
   Enter the power
   Press ENTER.
2. To find the root of a number (other than square root):
   On the TI-82:
   Enter the root index
   Press the key MATH
   With Math highlighted,
   Select $\sqrt[x]{}$
   Enter the number
   Press ENTER
   ❖ On TI-81, find roots by using fractional exponents.

   On the TI-85:
   Enter the root index
   Press the key MATH
   Select MISC
   Press MORE,
   From the secondary menu,
   Select $\sqrt[x]{}$
   Enter the number
   Press ENTER.
3. To find the factorial of a positive integer:
   On the TI-82:
   Enter the integer
   Press the key MATH
   Highlight PRB
   Select 4:!
   Press ENTER

   On the TI-85:
   Enter the integer
   Press the key MATH
   Select PROB
   From the secondary menu,
   select !
   Press ENTER.
4. To find the largest integer less than or equal to a given value (This is called the greatest integer function and is indicated by square brackets [ ]):
   On the TI-82:
   Press the key MATH
   Highlight NUM
   Select 4:Int
   Enter the value
   Press ENTER

   On the TI-85:
   Press the key MATH Select NUM
   From the secondary menu,
   select int
   Enter the value
   Press ENTER.
5. For the number π, press the key for π on the keyboard.
6. To find the absolute value of a given value:
   On the TI-82:
   Press the key ABS
   on the keyboard.

   On the TI-85:
   Press the key MATH
   From the menu, select NUM
   Select int
   Enter the value and press ENTER.

Procedure C6. Changing the mode of your calculator.

Press the key: MODE
You will see several lines. The items which are highlighted in each line set a particular part of the mode.

The first two lines set the way that numbers are displayed on the screen. In the first line, Norm is used for normal mode, Sci for scientific notation, and Eng for engineering notation. In the second line, Float causes numbers to be displayed in floating point form with variable number of decimal places and if one of the digits 0 to 9 is highlighted, all calculated numbers will be displayed with that number of decimal places.

The third line is used for trigonometric calculations and indicates if the angular measurements are in degrees or radians. Other lines on this screen will be considered later.

To change mode:
1. Use the cursor keys to move the cursor to the item desired.
2. Press ENTER. The new item should now be highlighted.
3. To return to the calculations screen, press the CLEAR or QUIT key.

(Example: To change the mode so that numbers are displayed in scientific notation with two decimal places, highlight Sci in the first line and press ENTER, then highlight 2 in the second line and press ENTER. Then, return to the calculations screen.)

Procedure C7. Storing a set of numbers as a list.

On the TI-82:
Up to six lists (of up to 99 values each) may be stored using the names L_1, L_2, L_3, L_4, L_5, and L_6

Method I:
1. Press the key {
2. Enter the numbers to be in the list separated by commas.
3. Press the key }
4. Press the key STO▷
5. Press the key for the list name and press ENTER.

On the TI-85:
Lists (of any length) may be stored using a list name of up to eight characters. The first character of the name a must be a letter.

Method I:
1. Press the key LIST.
2. Select {
3. Enter the numbers to be in the list separated by commas.
4. Select }
5. Press the key STO▷
6. Enter the list name and press ENTER.

Method II:
1. Press the key STAT
2. From the menu, select 1:EDIT
3. Move the highlight to the list and position where numbers are to be entered, and enter the number. The entered value will appear at the bottom of the screen. Press ENTER to place the number in the list.
4. After all values are entered, press QUIT to exit from the list table.

Note: A values may be changed by moving the highlight in the table to that value, entering the correct value, and pressing ENTER.

METHOD II:
1. Press the key LIST
2. Select EDIT
3. To enter a new list, enter the new name after Name= . To change values in a stored list enter the name of the list or select from the list of names.
4. Enter the values as elements of the list. After e1= enter the first value and press ENTER, after e2= enter the second value and press ENTER, etc. When finished, press QUIT.

Note: A value may be changed by moving the cursor to that value and entering the new value.

❖ Lists of numbers may not be stored on the TI-81.

<u>Procedure C8</u>. Viewing a set of numbers.

On the TI-82:
Press the key for the name of the list and press ENTER.
or
1. Press the key STAT
2. Select 1:EDIT
3. Move the highlight to the list.
4. Press QUIT to exit.

On the TI-85:
Enter the name of list and press ENTER.
or
1. Press the key LIST
2. Select NAMES, then select the list name and press ENTER or
   Select EDIT, enter or select list name, and press ENTER.

❖ Not available on the TI-81.

<u>Procedure C9</u>. To perform operations on lists.

1. To add, subtract, multiply or divide lists, enter the list or the name of a list, press the appropriate key (+, −, ×, or ÷), enter the second list or the list name and press ENTER. Lists must be of the same length in order for these operations to be performed.
2. To perform an operation on all the elements of a list, enter a value, press the appropriate key (+, −, ×, or ÷), enter the list or list name and press ENTER.
3. Functions such as finding the squares or square roots, may be performed on each element of a list by performing the function of the list or list name.

Note: For multiplication it is not necessary to use the symbol ×.

❖ Not available on the TI-81.

Procedure C10. To store a function.

On the TI-82:
The TI-82 may store up to eight functions, using function names $Y_1, Y_2, Y_3, \ldots. Y_8$.

Method I:
1. Enter inside of quotes the function to be stored.
2. Press the key STO▷
3. Press the key Y-VARS
4. Select 1:Function
5. Select a function name (This function will replace any previous function.)
6. Press ENTER

To see stored functions, press the key Y=.

Method II:
1. Press the key Y=
2. Enter the function to the right of a function name.
3. Press QUIT

To see a stored function, press the key Y=.

On the TI-85:
The TI-85 may store up to 99 functions using names that may contain up to eight characters. The first character of a name must be a letter.

Method I:
1. Enter the name of the function. (Names may be up to eight characters long.)
2. Press the key =
3. Enter the function.
4. Press ENTER

To see a stored function, press the key RCL, enter the name of the function, and press ENTER.

Method II:
(Stores functions under the names y1, y2, y3, ...)
1. Press the key GRAPH
2. Select: y(x)=
3. Enter the function after a function name.
4. Press QUIT

To see a stored function, press the key GRAPH, select y(x)= and press ENTER.

Note: Pressing CLEAR while the cursor is on the same line as the function will erase the function. Be sure to QUIT the function list before evaluating a function.

Procedure C11. To evaluate a function.

Functions may be evaluated at a single value or at several values by entering the values in a list.

On the TI-82:
1. Obtain the function name
   (a) Press the key Y-VARS
   (b) Select 1:Function...
   (c) Select the name of the function.

On the TI-85:
1. Obtain the function evalF
   (a) Press the key CALC
   (b) Select evalF
2. Enter either the function or the function name then a comma.

2. After the function name enter, in parentheses, the x value or a list of x values at which the function is to be evaluated. (Example: $Y_1(4)$ or $Y_1(\{1,2,3,4\})$)
3. Press ENTER

The results will be given as a value or a corresponding list. The TI-82 also has the ability to create a table of values for a function by pressing the key TABLE. Parameters for the table are set by pressing the key TblSet.

In TblSet:
TblMin is the value of the first x value in the table.
ΔTbl is the difference between table values.

3. Enter the variable for which the function is being evaluated followed by a comma.

The function names y1, y2, etc. may be entered from the keyboard (y must be lower case)
or

may be selected by:
(a) Pressing the key VARS
(b) From menu, select EQU (after pressing MORE).
(c) Select the variable name (may need page↓ or page↑).

4. Enter the value or a list of values at which the function is to be evaluated.
(Example: evalF(y1, x, 4) or evalF(y1,x,{1,2,4})
5. Press ENTER

Note: In the evaluation of a function, instead of placing a list of values inside of braces {}, we may enter the list name. See Procedure C7 for entering lists.

❖ On the TI-81, store a value for x, obtain the function name by pressing the key Y-VARS, press ENTER to evaluate the function.

Procedure C12. To store a constant for an alpha character.

1. Enter the value to be stored.
2. Press the key STO▷ (STO▷ shows as → on screen.)
3. Enter the alpha character.
4. Press ENTER

(Example: 6.35 → H)

Procedure C13. To obtain relation symbols (=, ≠, >, ≥, <, and ≤).

1. Press the key: TEST
2. Select the desired symbol from the menu.

Procedure C14. To change from radian mode to degree mode and vice versa.

1. Press the key: MODE
2. Move the cursor to either Radian or Degree on the third line depending on which mode is desired. (The mode which is highlighted is the active mode.) Press ENTER.
3. Press CLEAR or QUIT to return to the calculation screen.

Procedure C15. Conversion of Coordinates

Make sure calculator is in correct mode -- radians or degrees.

A. To change from polar to rectangular:

On the TI-82:

1. Press the key ANGLE
2. To find x:
   Select 7:P▷Rx(
   To find y:
   Select 8:P▷Ry(
3. Enter the values of r and θ separated by commas and press ENTER.

Example: P▷Rx(2.0, 45)
gives 1.4 [in degree mode]

On the TI-85:

1. Enter the magnitude and angle separated by a ∠ in [ ].
2. Press the key VECTR
3. Select OPS
4. Select ▷Rec (after pressing MORE)
5. Press ENTER

Example: [2.0∠45]▷Rec
gives [1.4, 1.4]

B. To change from rectangular to polar:

On the TI-82:

1. Press the key ANGLE
2. To find r:
   Select 5:R▷Pr(
   To find θ:
   Select 6:R▷Pθ(
3. Enter the values of x and y separated by commas and press ENTER.

Example: R▷Pθ(1.0, 1.0)
gives 45 [in degree mode]

Note: On the TI-82, it may be just as easy to use the conversion formulas (Equation 2) directly -- especially when converting from polar to rectangular.

On the TI-85:

1. Enter the values of x and y inside of [ ] separated by a comma.
2. Press the key VECTR
3. Select OPS
4. Select ▷Pol
5. Press ENTER

Example: [1.0, 1.0] ▷Pol
gives [1.4 ∠45]

Note: On the TI-85, vectors are entered or displayed as a pair of numbers inside of square brackets, [ ]. When the numbers are separated by a comma they are in rectangular form and when separated by ∠ are in polar form.

❖ On the TI-81: Press the key MATH, then:

A. For polar to rectangular, select 2:R▷R(and enter polar coordinates separated by a comma. Press ENTER to see the x-value and press Y to see the y-value.

B. For rectangular to polar, select 1:R▷P(and enter rectangular coordinates separated by a comma. Press ENTER to see the r-value and press θ to see the angle.

Procedure C16. Finding the sum of vectors in polar form.

On the TI-82:

1. Find the x- and y-components of each vector. (see Procedure C15)
2. Find the sum of the x-components and the sum of the y-components.
3. Find the magnitude and direction of the resultant vector. (see Procedure C15)

or

Find the sum

$A\cos\theta_A + B\cos\theta_B + \ldots$

and store this for x and find

$A\sin\theta_A + B\sin\theta_B + \ldots$

and store this for y, then convert [x, y] to polar form.

On the TI-85:

Vectors may be entered directly in polar form and the added together.

Example:

[12 ∠120] + [15 ∠75] gives [25 ∠95]

Note: On the TI-85, vectors may be stored under variable names by entering the vector, pressing the key STO▷, entering the variable name and pressing ENTER.

Vectors may also be entered or modified by first pressing the key VECTR, then selecting EDIT.

The sum of vectors may be found by finding the sum of the variables.

The form of the output of a vector on the screen may be changed by pressing the key MODE and in the seventh line highlighting

RectV for rectangular form

CylV for polar form

Procedure C17. Conversion of complex numbers.

Make sure calculator is in correct mode -- radians or degrees.

A. To change from polar to rectangular form:

On the TI-82:

Think of the complex number r ∠θ as r and θ in polar form and use Procedure C15.

On the TI-85:

1. Enter the magnitude and angle separated by a ∠ in ( ).
2. Press the key CPLX
3. Select ▷Rec (after pressing MORE) and press ENTER

Example: (2.0 ∠45)▷Rec gives (1.4, 1.4)

B. To change from rectangular to polar form:

On the TI-82:

Think of the complex number x + yj as x and y in rectangular form and use Procedure C15.

On the TI-85:

1. Enter the values of x and y inside of ( ) separated by a comma.
2. Press the key CPLX
3. Select ▷Pol
4. Press ENTER

Example: (1.0, 1.0) ▷Pol gives (1.4 ∠45)

Note: On the TI-85, a complex number is expressed as a pair of numbers inside of parentheses ( ). When the numbers are separated by a comma they are in rectangular form and when separated by ∠ they are in polar form.

Procedure C18. Operations on complex numbers.

On the TI-82:

1. If not already in rectangular form, change the complex number into rectangular form as when changing polar coordinates to rectangular coordinates (see Procedure C15).
2. Perform the operation on the rectangular forms of the complex numbers.
3. If the answer is to be in polar form, convert the result to polar form as when changing rectangular coordinates to polar coordinates (see Procedure C15).

Operations are performed by entering the complex numbers in either rectangular or polar form using the operation symbols +, −, ×, or ÷ between the numbers.

Note: Powers and roots of complex number may be found by raising the complex number to the desired power using ^.

The form of output of the complex number on the screen may be changed by pressing the key MODE and in the fourth line highlighting

RectC for rectangular mode

PolarC for polar mode

Procedure C19. To determine the numerical derivative from a function.

On the TI-82:

1. Press the key MATH
2. Select 8:nDeriv(

   The form is

   nDeriv(y, variable, value)

On the TI-85:

1. Press the key CALC
2. From the menu, select nDer

   The form is

   nDer(y, variable, value)

where y is the function itself or the name of the function, variable is the variable used in the function, and value is the number at which we wish to determine the derivative.

Example: nDeriv(Y_1,x,2)
The precision to which this is calculated may be changed by indicating the precision as a fourth value inside the parentheses. The value used otherwise is 0.001. (This value determines the Δx.)

Example: nDer(y1,x,2)
The precision to which this is calculated may be changed by pressing the key TOLER and changing the value for δ (this determines the Δx).

Note: The TI-85 also has der1 which determines the numerical first derivative as exactly as is possible and der2 which determines the numerical second derivative as exactly as possible.

❖ On the TI-81, the form is NDeriv(y, delx) where y is the function or name of the function and delx is a small value determining the accuracy of the calculation. Before using this, store the value for x at which the derivative is to be calculated.

<u>Procedure C20</u>. Function to calculate definite integral.

On the TI-82:
1. Press the key MATH
2. Select 9:fnInt(

On the TI-85:
1. Press the key CALC
2. Select fnint

The form of the function if fnInt(y, variable, a, b) where y is a function or name of a function, variable is the variable used in the function, and a and b are the lower and upper limits of integration.

(Example: fnInt(x^3, x, 1,3) is used to evaluate $\int_1^3 x^3\,dx$.)

❖ This function not available on the TI-81.

A.3 Graphing Procedures

Procedure G1. To enter and graph functions.

On the TI-82:

1. Press the key Y=
   This gives a screen showing a function table containing names of the functions:
   Y_1=
   Y_2=
   ...
2. Enter a function after the name of a function. Use the key X,T,θ for the variable x.
3. To graph functions, press the key GRAPH.
   To return to the calculations screen, press QUIT.

On the TI-85:

1. Press the key GRAPH
2. Select y(x)=
   This gives a screen showing a function table containing names of the functions:
   y1=
   ...
3. Enter a function after the name of a function. Use the key
   x-VAR for the variable x.
4. To graph functions, press EXIT to exit the secondary menu appearing on the screen, then select GRAPH from the menu.
   To return to the calculations screen, press QUIT.

Note:
1. Functions entered by this procedure will remain in the calculator's memory until changed or erased or until the memory is cleared.
2. While a graph is on the screen, pressing the key: CLEAR will clear the screen. However, the graph is still in the calculator's memory and can be seen again by pressing GRAPH.
3. When a function has been entered, the equal sign for that function becomes highlighted indicating that the function is turned ON - that is, it is an active function and will be graphed if the GRAPH key is pressed. Only functions which are turned ON will be graphed.
   To turn a function ON or OFF, with the function table on the screen:

   On the TI-82:
   (a) Move the cursor so that it falls on top of the equal sign of the function we wish to turn ON or OFF.
   (b) Press the ENTER key. If the function was ON it will be turned OFF and if it was OFF it will be turned ON.

   On the TI-85:
   (a) Move the cursor so it is on the same line as the function to be turned ON or OFF.
   (b) From the secondary, select SELCT. If the function was ON it will be turned OFF and if it was OFF it will be turned ON.

Procedure G2. To change or erase a function.

1. Display the function table on the screen (see Procedure G1).
2. (a) To change a function: Use the arrow keys to move the cursor to the desired location and make the changes by inserting, deleting, or changing the desired characters.
   (b) To erase a function: With the cursor on the same line as the function, press the key: CLEAR
3. Select GRAPH to graph the function or exit to the calculations screen by pressing QUIT.

Procedure G3. To obtain the standard viewing rectangle.

To obtain the standard viewing rectangle;

On the TI-82:
1. Press the key ZOOM
2. From the menu, select 6:ZStandard

On the TI-85:
1. Press the key GRAPH
2. From the menu, select ZOOM
3. From the secondary menu, select ZSTD.

For the standard viewing rectangle, the x-value at the left of the screen is −10 and at the right of the screen is 10 and the y-value at the bottom of the screen is −10 and at the top of the screen is 10. Each mark on the axes represents one unit.

Procedure G4. To use the trace function and find points on a graph.

1. Select the trace function:
   On the TI-82:
   Press the key TRACE

   On the TI-85:
   With the graph menu on the screen, select TRACE.
2. As the right and left arrow keys are pressed, the cursor moves along the graph and, at the same time, the coordinates of the cursor are given on the bottom of the screen.
3. If more than one graph is on the screen, pressing the up or down cursor keys will cause the cursor to jump from one graph to another. The number of the function will appear in the upper right of the screen.
4. If the graph goes off the top or bottom of the screen the cursor will continue to give coordinates of points on the graph.
5. If the cursor is moved off the right or left of the screen the graph scrolls (moves) right or left to keep the cursor on the screen.

Note: If the cursor is moved by using the cursor keys without first pressing the trace key, then the cursor can be moved to any point on the screen. The coordinates of this point will be given at the bottom of the screen.

Procedure G5. To see or change viewing rectangle.

1. To see values for the viewing rectangle on the screen.
   On the TI-82:
   Press the key WINDOW
   ❖ On TI-81, press RANGE
   On the TI-85:
   With the graph menu on the screen, select RANGE
2. To change viewing rectangle values:
   Make sure the cursor is on the quantity to be changed and enter a new value for that quantity. To keep a value and not change it, either (a) just press the key ENTER or (b) use the cursor keys to move the cursor to a value to be changed.
3. To see the graph, after the new values have been entered, select GRAPH
   To return to the calculations screen, press the key: QUIT

Note: The standard viewing rectangle has the following values:

Xmin = –10 Ymin = –10
Xmax = 10 Ymax = 10
Xscl = 1 Yscl = 1

Procedure G6: To obtain preset viewing rectangles.

1. Bring the zoom menu to the screen.
   On the TI-82:
   Press the key ZOOM
   On the TI-85:
   With the graph menu on the screen, select ZOOM
2. Select the desired preset viewing rectangle. (The left hand column gives the selection from the ZOOM menu for the TI-82 and the right hand column for the TI-85.)
   (a) For the standard viewing rectangle:
   Xmin = –10, Xmax = 10, Xscl = 1,
   Ymin = –10, Ymax = 10, Yscl = 1
   Select: 6:ZStandard ZSTD
   (b) For the square viewing rectangle:
   The square viewing rectangle keeps the y-scale the same and adjusts the x-scale so that one unit on x-axis equals one unit on y-axis. For a square viewing rectangle the ratio of y-axis to x-axis is about 2:3.
   Select: 5:ZSquare ZSQR (after pressing MORE)
   (c) For the decimal viewing rectangle:
   The decimal viewing rectangle makes each movement of the cursor (one pixel) equivalent to one-tenth of a unit
   Select: 4:ZDecimal ZDECM (after pressing MORE)
   (d) For the integer viewing rectangle:
   The integer viewing rectangle makes each movement of the cursor (one pixel) equivalent to one unit. After selecting the integer viewing rectangle, move the cursor to the point that is to be located at the center of the screen.

Select: 8:ZInteger ZINT (after pressing MORE)
Then press ENTER.
Note: To have one movement of the cursor differ by k units set:

Xmin = –94k/2	Xmin = –126k/2
Xmax = 94k/2	Xmax = 126k/2

(e) For the trigonometric viewing rectangle:
The trigonometric viewing rectangle sets the x-axis up in terms of π or degrees (depending on mode)and sets Xscl = $\frac{\pi}{2}$ (or 90°).
Select: 7:ZTrig ZTRIG

❖ The TI-81 does not have the decimal viewing rectangle.

Procedure G7. To graph functions on an interval.

1. To graph a function on the interval $x < a$ or on the interval $x \le a$ for some constant a:
   In the function table after the equal sign, enter the function f(x) followed by $(x<a)$ or $(x\le a)$.
   (Example: To graph $y = x^2$ on the interval $x<2$, enter for the function: $x^2(x<2)$)
2. To graph a function on the interval $a < x < b$ or on the interval $a \le x \le b$ for some constants a and b:
   In the function table after the equal sign, enter the function f(x) followed by $(x>a)(x<b)$ or $(x\le a)(x\le b)$.
   (Example: To graph $y = x^2$ on the interval $-3\le x\le 2$, enter for the function: $x^2(x\le -3)(x\le 2)$)
3. To graph a function on the interval $x > a$ or on the interval $x \ge a$ for some constant a:
   In the function table after the equal sign, enter the function f(x) followed by $(x>a)$ or $(x\ge a)$.
   (Example: To graph $y = x - 5$ on the interval $x>2$, enter for the function: $(x-5)(x>2)$)

Note: The above forms may be combined by writing their sum.

Procedure G8. Changing graphing modes.

1. To change modes:

On the TI-82:	On the TI-85:
Press the key MODE	With the graph menu on the screen, select FORMT (after pressing MORE)

   Then, use the cursor keys to highlight the desired option.
2. To draw graphs connected or with dots

Connected to connect plotted points with a line. Dot to just plot individual points.	DrawLine to connect plotted points with lines and DrawDot to just plot individual points.

3. To draw graphs simultaneously or sequentially

TI-82	TI-85
Sequential to graph each function in sequence or Simul to graph all selected functions at the same time.	SeqG to graph each function in sequence or SimulG to graph all selected functions at the same time..

4. Highlight the desired option in each line then press ENTER.
5. To return to the calculations screen, press CLEAR or QUIT.

Procedure G9. To zoom in or out on a section of a graph.

1. With a graph on the screen,

TI-82	TI-85
On the TI-82: Press the key ZOOM	On the TI-85: With the graph and graph menu on the screen, select ZOOM

2. From the menu:

TI-82	TI-85
To zoom in, select 2:Zoom In or To zoom in, select 3:Zoom Out	From the secondary menu, To zoom in, select ZIN or To zoom out, select ZOUT

3. The cursor keys may be used to move the cursor near the point of interest. After zooming, the point at the location of the cursor will be near the center of the screen.
4. Press the key: ENTER
5. After zooming, repeated zooms may be made (if no other key is pressed in the meantime) by moving the cursor near the desired point and pressing ENTER.

Note: It is a good idea to use the TRACE function together with ZOOM.

Procedure G10. To change zoom factors

1. Obtain the zoom menu:

TI-82	TI-85
On the TI-82: Press the key ZOOM	On the TI-85: With the graph menu on the screen, select ZOOM

2. Then

TI-82	TI-85
Highlight MEMORY in the top row and select 4:SetFactors.	From the secondary menu select ZFACT (after pressing MORE twice).

3. Change the factors as desired.
   There are factors for magnification in both the x- and y-directions. The preset factors are 4 in both directions. Make changes to XFact for the magnification in the x-direction and to YFact for magnification in the y-direction.
4. To exit this screen press the key: QUIT

Procedure G11. Solving an equation in one variable.

1. Write the equation so that it is in the form f(x) = 0, let y = f(x) and graph the function. Use a viewing rectangle so that the x-intercept of interest appears on the screen.
2. Determine a solution by:
   (a) Using Procedure G9 to zoom in on the x-intercept.
   or
   (b) Using the built in procedure:

   On the TI-82:
   (1) Press the key CALC
   (2) From the menu, select 2:root
   The words Lower Bound? appear.
   (3) Move the cursor near, but to the left of the intercept and press ENTER. The words Upper Bound? appear.
   (4) Move the cursor near, but to the right of the intercept and press ENTER. The word Guess? appears. It is sufficient at this point to just press ENTER again. The root (x-intercept) will appear at the bottom of the screen.

   On the TI-85:
   (1) From the graph menu, select MATH (after pressing MORE).
   (2) From the secondary menu, select ROOT.
   (3) Move the cursor near the intercept and press ENTER. The root (x-intercept) will appear at the bottom of the screen.

❖ Procedure 2(b) is not available on the TI-81. Use Procedure 2(a).

Procedure G12. Finding maximum and minimum points.

1. Graph the function and adjust the viewing rectangle so that the desired local maximum or local minimum point is on the screen.
2. Determine the local maximum or local minimum point by:
   (a) Using Procedure G9 to zoom in on the point
   or
   (b) Use the built-in procedure:

   On the TI-82:
   (1) Press the key CALC
   (2) From the menu, select 3:minimum or 4:maximum. The words Lower Bound? appear.

   On the TI-85:
   (1) Adjust the viewing rectangle so that the point of interest is the highest point or lowest point on the screen.

(3) Move the cursor near, but to the left of the desired point and press ENTER. The words Upper Bound? appear.
(4) Move the cursor near, but to the right of the desired point and press ENTER. The word Guess? appears. It is sufficient at this point to just press ENTER again. The coordinates of the point will appear at the bottom of the screen.

(2) From the graph menu, select MATH (after pressing MORE).
(3) From the secondary menu, select FMIN for a local minimum or FMAX for a local maximum point (after pressing MORE).
(4) Move the cursor near the desired point and press ENTER. The coordinates of the point will appear at the bottom of the screen.

Note: LOWER and UPPER may be used to select lower and upper bounds on which the maximum and minimum values are to be found.

❖ Procedure 2(b) is not available on the TI-81. Use Procedure 2(a).

Procedure G13. Find the value of a function at a given value of x.

1. Graph the function and adjust the viewing rectangle so that the given x-value is within the restricted domain of the screen.
2. Determine the value of the function by:
   (a) Using Procedure G9 to zoom in on the point
   or
   (b) Use the built-in procedure:

On the TI-82:
(1) Press the key CALC
(2) From the menu, select 1:value. The words Eval X= appear on the screen.
(3) Enter the given x-value and press ENTER. The value of the function will appear at the bottom of the screen.

On the TI-85:
(1) From the graph menu, select EVAL (after pressing MORE twice).
(2) The words Eval x = appear on the screen.
(3) Enter the x-value and press ENTER. The value of the function will appear at the bottom of the screen.

❖ Procedure 2(b) is not available on the TI-81. Use Procedure 2(a).

Procedure G14. To zoom in using a box.

1. Obtain the zoom menu:

On the TI-82:
Press the key ZOOM

On the TI-85:
With the graph menu on the screen, select ZOOM

2. Then
   Select 1:ZBox — From the secondary menu select BOX
3. Use the cursor keys to move the cursor to a location where one corner of the box is to be placed and press ENTER.
4. Use the cursor keys to move the cursor to the location for the opposite corner of the box and press ENTER. As the cursor keys are moved a box will be drawn and when ENTER is pressed, the area in the box is enlarged to fill the screen.

Procedure G15. Finding an intersection point of two graphs.

Enter the functions into the calculator and adjust the viewing rectangle so the region of the graph containing the intersection point on the screen. Then, either
Zoom in on the point of intersection. To obtain the desired degree of accuracy, use the trace function to move the cursor just to the left and then just to the right of the intersection point after each zoom. When the x-values, on either side of the intersection point, rounded off to the desired number of significant digits are equal, we have reached a solution. (The y-values may also be checked by moving the cursor just above and just below the intersection point.)

❖ Use the above procedure on the TI-81.

or

Determine the point of intersection by the built-in procedure:

1. Select the correct mode to find the point of intersection.

On the TI-82:	On the TI-85:
Press the key CALC Select 5:intersect	With the graph menu on the screen, select MATH (after pressing MORE). Then, select ISECT (after pressing MORE).

2. Move the cursor near the point of intersection.
3. If the cursor is on one of the graphs, continue. Otherwise, use the up and down cursor keys to move the cursor to the correct graph. (On the TI-82 the words "First curve?" appear on the screen.)
4. Press ENTER.
5. If the cursor is on the correct second graph, continue. Otherwise, use the up and down cursor keys to change the cursor to the correct graph. (On the TI-82 the words "Second curve?" appear.)
6. Press ENTER.

7. Obtain the coordinates of the intersection point.

   On the TI-82:
   The word "Guess?" appears on the screen. Make sure the cursor is near the point of intersection and press ENTER.
   The word Intersection and the coordinates of the intersection point will be given at the bottom of the screen.

   On the TI-85:
   The word ISECT and the coordinates of the intersection point will be given at the bottom of the screen.

Procedure G16. To graph each of two functions and then their sum.

1. Set the correct values for the viewing rectangle.
2. Enter the first function into the function table for Y_1.
3. Enter the second function into the function table for Y_2.
4. Move the cursor after the equal sign for Y_3. Press the Y-VARS key and select Y_1 from the menu. (See Procedure C11)
5. Press the key: +
6. Press the Y-VARS key and select Y_2 from the menu. (The third line should now appear as $:Y_3=Y_1+Y_2$)
7. Press GRAPH

Procedure G17. To change graphing modes for regular functions, parametric, or polar equations.

1. Set the graphing mode:
   Press the key MODE.
   Make active:

On the TI-82:	On the TI-85:
(a) For graphing in rectangular coordinates:	
In fourth line: Func	In fourth line: RectC In fifth line: Func
❖ On TI-81, make active Function and Rect.	
(b) For graphing parametric equations:	
In fourth line: Par	In fourth line: RectC In fifth line: Param
❖ On TI-81, make active Param and Rect	
(c) For graphing in polar coordinates:	
In fourth line: Pol	In fourth line: PolarC In fifth line: Pol
❖ On TI-81, make active Param and Polar	

   Then, press QUIT or CLEAR.

2. Set the graphing format:
   ❖ Not available on TI-81.

   On the TI-82:
   Press the key WINDOW
   Highlight FORMAT in the top row.
   Then in second line:

   On the TI-85:
   Press the key GRAPH
   From the menu, select FORMT (after pressing MORE)
   Then in first line:

   Make active:
   (a) For graphing in rectangular coordinates or for parametric equations: RectGC
   (b) For graphing in polar coordinates: PolarGC

   Then, press QUIT or CLEAR.

Note: To make an item active, move the cursor to that item and press ENTER.

Procedure G18. To graph parametric equations.

1. Make sure the calculator is in correct mode for parametric equations. (Procedure G17)
2. To enter the equations:

   On the TI-82:
   Press the key Y=
   On the screen will appear:
   $X_{1t}=$
   $Y_{1t}=$
   $X_{2t}=$
   $Y_{2t}=$
   (etc.)
   (up to 6 pairs of equations may be entered)

   On the TI-85:
   Press the key GRAPH.
   From the menu, select E(t) =
   On the screen will appear:
   xt1=
   yt1=
   (up to 99 pairs of equations may be entered.)

   Functions are not turned on (the equal signs are not highlighted) until functions have been entered for both x and y. Functions may be turned on and off as with functions in rectangular coordinates (see Note 3, Procedure G1). Only functions turned on will be graphed.
3. Enter the first equation for x in the first row and the corresponding equation of y in the second row. If there are other sets of functions you wish to graph, enter them for the other x's and y's.
   To obtain the variable t

   On the TI-82:
   Press the key X,T,θ

   On the TI-85:
   Select t from the menu.
4. Press or select GRAPH to see the graph or QUIT to exit.

Procedure G19. To change values for the parameter T.

With the calculator in the mode for parametric equations:
1. Follow Procedure G5 to see the values for the viewing rectangle.
2. Enter new values for Tmin, Tmax, and Tstep and press ENTER or move to the next item by using the cursor keys.

The standard viewing rectangle values are:

Tmin = 0, Tmax = 6.28...(2π) , and Tstep = 0.13...(π/24)
(or Tmax = 360 and Tstep = 7.5 if in degree mode).

Tstep determines how often values for x and y are calculated. The size of Tstep will affect the appearance of the graph. Tmax should be sufficiently large to give a complete graph.

3. Press QUIT to exit.

Procedure G20. To graph equations involving y^2.

1. Solve the equation for y. There will be two functions: y = f(x) and y = g(x). In many cases it will be true that g(x) = −f(x).
2. Store f(x) under one function name and g(x) for the second function name (see Procedure C10).
   (Example: Y_1 = f(x)
   Y_2 = g(x))
   or
   If g(x) = −f(x), we may do the following:
   (a) Store f(x) after a function name (Example: Y_1).
   (b) Move the cursor after the equal sign for a second function name (Example: Y_2).
   (c) Enter the negative sign: (−)
   (d) Obtain and enter the name of the first function.
   (Example: $Y_2 = -Y_1$) (See Procedure C11.)
3. Graph the functions.

Procedure G21. To obtain special Y-variables.

On the TI-82:

1. To obtain Ymin or Ymax
   Press the key: VARS
   From the menu,
   select: Window..
   ❖ On TI-81, select RNG
   Select Ymin or Ymax
2. To obtain variable names
   Y_1, Y_2, etc.
   Press the key: Y-VARS
   From the menu, select:
   Function
   Select the function name.

On the TI-85:

The best way is perhaps to just enter the name from the keyboard. When entering the name from the keyboard, be sure the correct upper or lower case is entered.

or

1. Press the key VARS
2. Select ALL
3. Press F1 to page down till desired variable is on screen.
4. Use cursor keys to select desired variable.
5. Press ENTER

Procedure G22. To shade a region of a graph.

1. Enter the desired function for a function name. Make sure all functions not wanted are deleted or turned off.
2. Select appropriate window values and graph the function.
3. Select the Shade command:

   On the TI-82:
   Press the key: DRAW
   From the menu, select 7:Shade(
   Form is
   Shade(L, U, D, Lt, Rt)

   On the TI-85:
   From the GRAPH menu, select DRAW (after pressing MORE) and select Shade
   Form is Shade(L, U, Lt, Rt)

   Where:
   L is lower boundary (value or function) of the shaded area.
   U is upper boundary (value or function) of the shaded area.
   D is the density value (a digit from 1 to 8) that determines the spacing of the vertical lines the calculator draws to shade the region. The higher the value for the density, the more widely spaced the lines. If the density value is omitted, the shading is solid.
   Lt is the left most x-value.
   Rt is the right most x-value.
   [D, Lt, Rt are optional - but on the TI-82, if Lt and Rt at included, D must be also included.]
   (Example: Shade(-10,Y_1) shades in the area above $y = -10$ and below the graph of Y_1)
4. Enter the appropriate values for L, U, and/or D and left and right x-values. Then, press ENTER.

Note: To shade above the function represented by Y_1, use Shade(Y_1, Ymax).
To shade below the function represented by Y_1, use Shade(Ymin, Y_1).

Procedure G23. To graph in polar coordinates

1. Make sure the calculator is in correct mode for polar coordinates. (Procedure G17)
2. To enter polar equations:

   On the TI-82:
   Press the key Y=
   On the screen will appear:
   r1=
   r2=
   r3=
   (etc.)
   (up to 6 equations may be entered)

   On the TI-85:
   Press the key GRAPH.
   From the menu, select r(θ) =
   On the screen will appear: r1=
   (up to 99 polar equations may be entered as r1, r2, r3, ...)

   Functions may be turned on and off as with functions in rectangular coordinates (see Procedure G1, Note 3).

3. Enter the equation for r.
   For the variable θ
   On the TI-82:
   Press the key X,T,θ
   On the TI-85:
   Select θ from the menu.
4. Press or select GRAPH to graph the function or QUIT to exit.

❖ The TI-81 does not have direct polar graphing. To graph $r = f(\theta)$ in polar coordinates, graph the parametric equations $x = f(T) \cos T$, $y = f(T) \sin T$ after setting the calculator for polar coordinate graphing. T is used for θ.

Procedure G24. To change viewing rectangles values for θ.

With the calculator in the mode for polar coordinates:

1. Follow Procedure G5 to see the values for the viewing rectangle.
2. Enter new values for θmin, θmax, and θstep and press ENTER or move to the next item by using the cursor keys.
   The standard viewing rectangle values are

   θmin = 0, θmax = 6.28... (2π), and θstep = 0.13...(π/24)
   (or θmax = 360 and θstep = 7.5 if in degree mode).

   θstep determines how often points are calculated and may affect the appearance of the graph. Leaving θstep at approximately 0.1 is generally sufficient but θstep may have to be changed if there is a major change in the viewing rectangle. θmax should be sufficiently large to give a complete graph.
3. Press QUIT to exit.

❖ For TI-81, see Procedure G19

Procedure G25. Drawing a tangent line at a point.

1. Store the function for a function name and graph the function.
2. Draw the tangent line.

On the TI-82:

(a) While on the calculations screen, press the key DRAW.
(b) Select: 5:Tangent(
    The form is
    Tangent(y, value)
    Where y is the function or function name and value is the x-value.
    Example: Tangent(Y_1,2)
(c) Press ENTER

On the TI-85:

(a) With the graph on the screen select DRAW from the GRAPH menu (after pressing MORE)
(b) From the secondary menu, select TanLn (after pressing MORE twice). The form is
    TanLn(y, value).
    where y is the function or function name and value is the x-value.
    Example: TanLn(y1,3)

The tangent line may also be drawn by:
(a) With the graph on the screen and the cursor at the correct point, press DRAW.
(b) Select 5:Tangent(and press ENTER to return to graph.
(c) Press ENTER a second time. The tangent line will appear on the graph and the slope of the line will be given at the bottom of the screen.

The tangent line may also be drawn by:
(a) With the graph and the GRAPH menu on the screen, select MATH (after pressing MORE).
(b) From the secondary menu, select TANLN (after pressing MORE twice).
(c) Move the cursor to the desired point and press ENTER. The tangent line will be drawn on the graph and the it slope given at the bottom of the screen.

❖ This feature not available on the TI-81.

Procedure G26. Determining the value of the derivative from points on a graph.

On the TI-82:
1. With a graph on the screen, press the key CALC.
2. Select 6:dy/dx.
3. Move the cursor to the desired point and press ENTER. The value of the numerical derivative appears on the bottom of the screen.

On the TI-85:
1. With the and the graph menu on the screen, press the key MATH (after pressing MORE).
2. From the secondary menu select dy/dx.
3. Move the cursor to the desired point and press ENTER. The value of the numerical derivative appears at the bottom of the screen.

❖ This feature not available on the TI-81.

Procedure G27. Function to determine the area under a graph.

With the graph of a function on the screen:

On the TI-82:
1. Press the key CALC
2. Select 7:∫f(x)dx
3. Move the cursor to the first x-value and press ENTER.
4. Move the cursor to the second x-value and press ENTER.
(The area will be shaded.)

On the TI-85:
1. Select MATH (after pressing MORE)
2. Select ∫f(x)
3. Move the cursor to first x-value and press ENTER.
4. Move the cursor to the second x-value and press ENTER.

The area between the curve, the x-axis, and the two x-values will be given at the bottom of the screen.

❖ This feature not available on the TI-81.

A.4 Programming Procedures

Procedure P1. To enter a new program.

1. Press the key: PRGM
2. Enter programming mode:

On the TI-82:	On the TI-85:
Highlight NEW in the top row and select 1:Create New	Select EDIT.

   ❖ On the TI-81, highlight EDIT and select a program.
3. After Name=, enter a name for the program. The name may be up to eight characters long. (Notice that the calculator is ready for alpha characters.) Then, press ENTER.

   ❖ On the TI-81, enter a name after program number (Prog1).
4. The word PROGRAM appears on the top row followed by the name of the program and the cursor is on the second row preceded by a colon (:). The calculator is now in programming mode and is ready for the first statement in the program.
5. Enter the steps of the program one line at a time, pressing ENTER after each step.
6. When finished entering the program, press QUIT to return to the calculations screen.

Procedure P2. To execute, edit, or erase a program.

1. Press the key: PRGM
2. To execute a program:

On the TI-82:	On the TI-85:
With EXEC highlighted in the top row, select the name of the program to be executed. Then, press ENTER..	Select NAMES From the secondary menu, select the name of the program. Then, press ENTER.

   A program may be executed additional times by pressing ENTER if no other keys have been pressed in the meantime.
3. To edit a program:

On the TI-82:	On the TI-85:
Highlight EDIT in the top row and select the name of the program to be edited.	Select EDIT. From the secondary menu, select the name of the program to be edited.

   Make changes in the program as desired. The cursor keys may be used to move the cursor to any place in the program. The calculator is now in programming mode.
4. To erase a program:

On the TI-82:	On the TI-85:
Press the key MEM. Select 2:Delete... From the secondary menu, select 6:Prgm...	Press the key MEM Select DELET Select PRGM (after pressing MORE)

Move the marker on the left until it is next to the name of the program to be erased and press ENTER. The program will be erased from memory.

❖ On the TI-81, highlight ERASE, select the program to be erased, and press ENTER.

Procedure P3. To input a value for a variable.

Entering a value for a variable uses the keyword Input.
The form of the statement is: **Input v** (v represents the variable name)

With the calculator in programming mode:

1. To access the word Input:

   On the TI-82:
   Press the key PRGM
   Highlight I/O in the top row.
   Select 1:Input.

   On the TI-85:
   From the menu, select I/O
   From the secondary menu, select Input

2. Enter the variable name and press ENTER.
   (Example: Input C)

When the program is executed, the word Input followed by a variable will cause a question mark to appear on the screen. Any number entered is stored for that variable and when the variable is later used in the program that value is used in the calculation.

Procedure P4. To evaluate an expression and store that value to a variable.

With the calculator in programming mode:

On the TI-82:
Enter the expression, press the key STO▷, and then, enter a variable name.

On the TI-85:
Enter the expression, press the key STO▷ and then, enter a variable name.
or
Enter a variable name, press the key =, and then enter the expression.

(Example: $\frac{9}{5}C + 32 \rightarrow F$)

Procedure P5. To display quantities from a program.

Either the value for a variable or an alpha expression may be displayed by using the keyword Disp.
The form of the statement is: **Disp v** (v represents the variable name)
The quantity v may be a variable name, a number, or a set of characters enclosed inside of quotes. If v is a quantity enclosed in quotes, the quantity will be printed exactly as it appears.

With the calculator in programming mode:

1. To access the word Disp:

On the TI-82:	On the TI-85:
Press the key PRGM	From the menu, select I/O
Highlight I/O in the top row.	From the secondary menu,
Select 3:Disp.	select Disp

2. Enter the variable name and press ENTER. (Example: Disp F)

When the program is executed, the value stored for the variable or the expression will be displayed on the screen.

Procedure P6. To add or delete line from a program.

With the calculator in programming mode:

1. To add a line:
   (a) To insert a line before a given line, move the cursor to the first character of the line and to insert a line after a given line, move the character to the last character in the given line.
   (b) Press the key INS, then press ENTER This will create a blank line either before or after the given line
   (c) Make sure the cursor is in the desired line. An instruction may now be entered.
2. To delete a line
   (a) Move the cursor to the desired line
   (b) Press the key CLEAR, then press DEL.

Procedure P7. Finding the area under a graph.

1. The desired function should be stored under the function name in the function table. (Procedure C6)
2. Enter the following program into your calculator to find the area between two values B and C: (It is assumed that the function is stored under the function name Y_1)

```
PROGRAM:AREA
Disp "ENTER FIRST Z"
Input B
Disp "ENTER SECOND Z"
Input C
fnInt(Y1,X,B,C)→I
Disp "AREA IS:"
Disp I
```

To obtain the definite integral function fnInt:

On the TI-82:	On the TI-85:
Press the key MATH	Press the key CALC
From the menu, select 9:fnInt(	From the menu, select fnInt

The form of the definite integral function is

fnInt(function name, variable, smaller value, larger value)

(The function may also be used without being in a program.)

❖ The fnInt function is not available on the TI-81. It will be necessary to use the Trapezoidal or Simpson's Rule -- see Section 14.3.

Procedure P8. To display a graph from within a program.

While in programming edit mode (at a step in program):

On the TI-82:

1. Press the key PRGM
2. Select I/O
3. Select 4:DispGraph

On the TI-85:

1. Select I/O
2. Select DispG

Procedure P9. Creating a loop (Lbl, If, and Goto commands).

While in programming edit mode (at a step in program):

On the TI-82:

1. Select the label (Lbl) command:
   a. Press PRGM
   b. With CTL highlighted, select 9:Lbl
2. Select the If command:
   a. Press PRGM
   b. With CTL highlighted, select 1:If
3. Select the Goto command:
   a. Press PRGM
   b. With CTL highlighted, select 0:Goto

Note: A label may be a letter or a numerical digit.

On the TI-85:

1. Select the label (Lbl) command:
   a. Select CTL
   b. Select Lbl (after pressing MORE)
2. Select If command
   a. Select CTL
   b. Select If
3. Select the Goto command
   a. Select CTL
   b. Select Goto (after pressing MORE)

Note: A label may be a name of up to 8 characters, beginning with a letter.

The form of the label command is Lbl *label*
The form of the If command is If *condition*
The form of the Goto command is Goto *label*
The construction of the loop is of the form:

... (program statements before loop)
Lbl *label*
... (program statements within loop)
...
If *condition*
Goto *label*
... (program statements after loop)

Procedure P10. Drawing a line on the screen from within a program (Line command).

While in programming edit mode (at a step in program):

On the TI-82:
1. Press the key DRAW
2. With DRAW highlighted, select 2:LINE

On the TI-85:
1. Press the key GRAPH
2. Select DRAW (after pressing MORE)
3. Select LINE

The form of the line command is LINE(*a, b, c, d*) where a and b are the coordinates of one point and c and d are the coordinates of a second point. This command draws a line from (a, b) to (c, d).
Example: LINE $(0,2,X,Y_1)$ draws a line from the point (0,2) to the point (x,f(x)).

Procedure P11. To temporarily halt execution of a program. (PAUSE statement)

While in programming edit mode (at a step in program):

On the TI-82:
1. Press the key PRGM
2. With CTL highlighted, select 8:PAUSE

On the TI-85:
1. Select CTL
2. Select Pause (after pressing MORE twice)

The PAUSE statement when used in a program halts execution of the steps of a program until the ENTER key is pressed.

A.5 Matrix Procedures

Procedure M1. To enter or modify a matrix.

1. Press the key MATRX. This gives the matrix menu.
2. Enter EDIT mode.

   On the TI-82:
   Highlight EDIT in the top row.

   On the TI-85:
   Select EDIT from the menu.
3. Select a name for the matrix.

   On the TI-82:
   Select one of the five matrices [A], [B], [C], [D], or [E]. This will be the new matrix or the matrix to be edited.

   On the TI-85:
   Enter the name for a new matrix or select the name of a matrix to be edited, then press ENTER. The name may be up to eight characters long (only 5 letters will show in name box).
4. The first line on the screen will contain the word MATRIX followed by the name of the matrix and two numbers separated by ×. These two numbers represent the size of the matrix. First enter the number of rows and press ENTER, then the numbers of columns and press

ENTER. To keep the values the same, use the cursor keys to move the cursor or just press ENTER.

5. Enter the elements of the matrix.

   On the TI-82:
   The element of the matrix to be entered or changed is highlighted and at the bottom of the screen is the row and column number of the element followed by its value.
   As a value is entered, the number at the bottom of the screen will change. To enter this value into the matrix, press ENTER. Values are entered by rows. Continue until all elements are correct.
   The cursor keys are used to move the highlight to elements to be changed.

   On the TI-85:
   On the left of the screen are the row and column numbers of elements of the matrix. A single column appears on the screen at a time. After each value is entered, press ENTER to go to the next value. Values are entered by row. (As each value is entered the calculator will jump to the next column.) The cursor keys may be used to go from one element to the next if an element is not to be changed. The menu selections ◁Col and Col▷ may be used to change columns.

6. Exit from matrix edit mode by pressing the key: QUIT. (Pressing CLEAR will clear the value for a given element.)

❖ On the TI-81, elements of a matrix are enter after the position numbers that appear on the left of the screen. Elements are entered row by row.

Procedure M2. To select the name of a matrix and to view the matrix.

1. Press the key MATRIX. This gives the matrix menu.
2. Select the name of the matrix.

   On the TI-82:
   With NAMES highlighted in the top row, select one of the five matrices [A], [B], [C], [D], or [E].

   On the TI-85:
   Select NAMES from the matrix menu. From the secondary menu, select the name of the matrix.

3. View the matrix by pressing ENTER. In some cases the matrix may extend off the screen to the right or to the left. To see that part of the matrix not on the screen use the right or left cursor keys to scroll (move) the unseen part onto the screen.

❖ On the TI-81, names of matrices are selected by pressing one of the keys labeled [A], [B], or [C]. Matrices may be viewed by pressing ENTER after selection of a name.

Procedure M3. Addition, subtraction, scalar multiplication or multiplication of matrices.

See Procedure M2 for how to select the names of matrices.

1. Addition of two matrices:
   Select the name of the first matrix, press the addition sign, select the

name of the second matrix, then press ENTER.
(Example: [M] + [N])

2. Subtraction of two matrices:
   Select the name of the first matrix, press the subtraction sign, select the name of the second matrix, then press ENTER.
   (Example: [M] – [N])
3. Multiplication of matrix by a scalar:
   Enter the scalar, select the name of the matrix and press ENTER.
   (Example: 2.5 [M] or 2.5×[M])
4. Multiplication of two matrices:
   Select the name of the first matrix, select the name of the second matrix, and press ENTER.
   (Example: [M][N] or [M]×[N])

Many of these operations may be combined into one statement. The result of a matrix calculation is stored under ANS.

❖ On the TI-81, several operations may not be combined due to memory restrictions. After performing one operation the resulting matrix is stored in ANS. This matrix may then be used in additional operations.

Procedure M4. To find the determinant, inverse, and transpose of a matrix.

1 To find the determinant:

On the TI-82:
Press the key MATRX
Highlight MATH in top row.
From the menu, select det.

On the TI-85:
From the matrix menu, select MATH.

Then select the name of the matrix (Procedure M2) and press ENTER.
(Example: det[A] or det A}

2. To find the inverse:
   Select the name of the matrix.
   Press the key x^{-1} and press ENTER.
   (Example: $[A]^{-1}$ or A^{-1})
3. To find the transpose:
   Select the name of the matrix, then

   On the TI-82:
   Press the key MATRX
   Highlight MATH in top row.

   On the TI-85:
   From the matrix menu, select MATH.

   From the menu, select ^{T} and press ENTER.
   (Example: $[A]^T$ or A^T)

❖ On the TI-81, highlight MATRIX in the top row.

Procedure M5. To store a matrix.

1. Select the name of the first matrix or enter a matrix expression.
2. Press the key STO▷
3. Select the name of the second matrix and press ENTER.
   (Example: [A] + [B] → [C] or A + B →C)
   On the TI-85, we may also use an equal sign: C = A + B.

Note: Storing a matrix for a second matrix, erases the previous contents of the second matrix.

Procedure M6: Row operations on a matrix.

1. Obtain the matrix operations menu by first pressing the key MATRX, then:

On the TI-82:	On the TI-85:
Highlight MATH in top row.	From the matrix menu, select OPS
❖ On the TI-81, highlight MATRIX in top row.	(then press MORE).

2. Select the desired operation from the menu. Names of matrices are selected as in Procedure M1. (The left hand column gives the selection, form, and example for the TI-82 and the right hand column given the selection, form, and example for the TI-85.)

   (a) To interchange two rows on a matrix, select

TI-82	TI-85
rowSwap(	rSwap
Form:	Form:
rowSwap(name, row1, row2)	rSwap(name, row1, row2)
Example:	Example:
rowSwap([A], 1, 3)	rSwap(A, 1, 3)

   (Interchanges rows 1 and 3 of matrix A.)

   (b) To multiply a row by a constant, select:

TI-82	TI-85
*row(	multR
Form:	Form:
*row(multiplier, name, row)	multR(multiplier, name, row)
Example:	Example:
*row(1/2, [A], 3)	multR(1/2, A, 3)

   (Multiplies row 3 of matrix A by ½.)
   To divide a row by a constant, multiply by the reciprocal.

   (c) To multiply a row by a constant and add to another row, select

TI-82	TI-85
*row+(	mRAdd
Form:	Form:
*row+(mult,name,row1,row2)	mRAdd(mult,name,row1,row2)
Example:	Example:
*row+(−5,[A], 1, 3)	mRAdd(−5, A, 1, 3)

   (Multiplies row 1 of matrix A by −5 and adds the result to row 3 and replaces row 3.)

3. Press ENTER.

A.6 Statistical Procedures

Procedure S1. To enter or change single variable statistical data.

On the TI-82 and TI-85 calculators data is stored as lists of numbers. (See also Procedure C7)

1. Press the key STAT
2. Enter edit mode.
   On the TI-82:
   With EDIT highlighted in the top row, select 1:Edit...
   ❖ On the TI-81, highlight DATA in top row.
   On the TI-85:
   Select EDIT from the menu.
3. Enter the data values and frequencies.
   All frequencies must be entered as integers.
   On the TI-82:
   Enter the data values under one list name. After entering every data value, press ENTER.
   Enter the related frequencies under a second list name. After entering each frequency, press ENTER. Make sure the frequencies are entered in the same order as the data items so that they correspond to the correct item.
   On the TI-85:
   The calculator will ask for an xlist name and a ylist name. The xlist is for the data values and the ylist is for the frequencies.
   Enter a name (up to eight characters) for xlist and press ENTER and enter a name for ylist and press ENTER.
   Enter each data value as an x-value and the corresponding frequency as a y-value. (y_1 is the frequency for data item x_1, y_2 for x_2, etc.)
   Note: When there are intermediate data values with a frequency of zero, it is best, for graphing purposes, to enter these data values and the zero frequencies.
   ❖ On the TI-81, data values are entered as x-values and frequencies are entered as y values.
4. To edit data.
   On the TI-82:
   Move the cursor to the values to be changed, make the change, and press ENTER. As the data item in a list is highlighted, the value of that item appears at the bottom of the screen.
   On the TI-85:
   Move the cursor to the values to be changed and make the desired change.
5. To exit, after all data have been entered and the cursor appears on the next value, press QUIT.
6. To view a list, see Procedure C8.

Procedure S2. To graph single variable statistical data.

1. Make sure the viewing rectangle is appropriate for the data to be graphed, that the graph display has been cleared, and that all functions in the function table have been turned off.
2. Be sure the data to be graphed is stored in lists in the calculator. (Procedure S1)
3. Define the plot and draw the graph:

On the TI-82:

(a) Press the key STAT PLOT

(b) Select one of the three plots.

The screen will show the plot number selected, followed by several options. The highlighted options are active. To highlight an option, move the cursor to that option and press ENTER.

(c) Highlight ON.

(d) Highlight the type of graph. The first symbol following the word Type is for a scatter plot, the second for a xyLine, the third for a box plot, and the fourth for a histogram.

(e) For Xlist, highlight the name of the list of data values.

(f) For Ylist, highlight the name of the list of frequencies.

(g) For Mark, highlight the symbol to be used in plotting data.

(h) To exit, press QUIT

(i) To see the graph, press GRAPH.

Turn the STAT PLOT off when finished.

On the TI-85:

Press the key STAT

From the menu, select DRAW

From the secondary menu, to draw a graph, select:

HIST for a histogram.

SCAT for a scatter graph

xyLine for a frequency polygon.

Note: For a histogram the value of Xscl will determine width of bars.

One type of graph should be cleared before displaying another.

❖ On the TI-81, press the key STAT and highlight DRAW in top row. Select the type of graph and press ENTER.

Procedure S3. To clear the graphics screen.

1. All functions should be deleted or turned off.
2. Clear the screen:

On the TI-82:
Press the key DRAW.
With DRAW highlighted in the first row, select 1:ClrDraw.
Press ENTER.

On the TI-85:
Press the key STAT
From the menu, select DRAW.
From the secondary menu, select CLDRW.

Procedure S4. To obtain mean, standard deviation and other statistical information.

Be sure one-variable data has been stored as in Procedure S1.

1. Press the key STAT
2. Obtain the statistical information:

On the TI-82:
(a) Highlight CALC in the top row.
(b) From the menu, select 3:SetUp
(c) Under 1-Var Stats, select the list name to be used for the data values and the list name to be used for the frequencies. (If each has a frequency of one, select 1.)
(d) Press QUIT
(e) Again press STAT and highlight CALC in the top row.
(f) From the menu, select 1:1-Var Stats, then press ENTER

On the TI-85:
(a) From the menu, select CALC.
(b) Make sure the xlist name is correct and press ENTER and the ylist name is correct and press ENTER.
(c) From the secondary menu, select 1-VAR.

❖ On the TI-81, select 1-VAR and press ENTER.

A list of statistical quantities appear.

$\overline{x}$ is the mean of all the x-values
$\sum x$ is the sum of all the x-values
$\sum x^2$ is the sum of all the squares of the x-values
Sx is the s-standard deviation of the x-values
σx is the σ-standard deviation of the x-values
n is the total number of x-values

5. Clear the screen.

On the TI-82:
Press CLEAR to clear the screen.

On the TI-85:
Press QUIT to exit.

Procedure S5. To clear statistical data from memory.

1. Press the key: MEM.
2. From the menu, select Delete.
3. Then select, List.
4. Use the cursor keys to move the indicator to the name of the list to be deleted and press ENTER.
5. Press QUIT to exit.

Note: On the TI-82, data may also be deleted by: Press the key STAT, select 4:ClrList, enter the names, separated by commas, of the lists to be deleted (by pressing the keys for those names), then press ENTER.

❖ On the TI-81, press STAT, highlight DATA in the top row, and select 2:CllrStat. Then, press ENTER.

Procedure S6. To enter or change two variable statistical data.

On the TI-82 and TI-85 calculators, data is stored as lists of numbers. (See Procedure C7)

1. First press the STAT.
2. Enter edit mode.

   On the TI-82:
   With EDIT highlighted in the top row, select 1:Edit...

   ❖ On the TI-81, highlight DATA in the top row.

   On the TI-85:
   Select EDIT from the menu.
3. Enter the x-values as one list and the y-values as a second list:

   On the TI-82:
   Enter the x-values under a list name. After every value is entered, press ENTER.
   Enter the related y-values under a second list name. After each value is entered, press ENTER.
   Make sure the y-values are entered in the same order as the x-values so that they correspond.

   On the TI-85:
   The calculator will ask for an xlist name and a ylist name. The xlist is for the x-values and the ylist is for the y-values. Enter a name (up to eight characters) for xlist and press ENTER and enter a name for ylist and press ENTER .
   Enter the first x-value and the corresponding y-value, the second x-value and the second y-value, etc.

   ❖ On the TI-81, enter x- and y-values for each point. Frequencies of points can not be entered.

4. To edit data.

   On the TI-82:
   Move the cursor to the values to be changed, make the change, and press ENTER. As the data item in a list is highlighted, the value of that item appears at the bottom of the screen.

   On the TI-85:
   Move the cursor to the values to be changed and make the desired change.
5. To exit, after all data have been entered and the cursor appears on the next value, press QUIT.
6. To view a list, see Procedure C8.

Procedure S7. To determine a regression equation.

1. Be sure two-variable data has been stored as in Procedure S6.
2. Obtain the proper menu:

   On the TI-82:
   (a) Press the key STAT
   (b) Highlight CALC in the top row.
   (c) From the menu, select 3:SetUp.
   (d) Under 2-Var Stats highlight the list name to be used for the x-values and the list name to be used for the y-values. For Freq highlight 1 unless the number of times various points occur are listed under a third list name.
   (e) Press QUIT.
   (f) Again press the key STAT and highlight CALC in the top row.

   On the TI-85:
   (a) Press the key STAT
   (b) From the menu, select CALC.
   (c) Make sure the xlist name is correct and press ENTER and the ylist name is correct and press ENTER.

   ❖ On the TI-81, with CALC highlighted in the top row, select the desired type of equation.
3. Select the type of regression equation. (The form of the menu item for the TI-82 is on the left and for the TI-85 is on the right.)
   (a) For linear regression ($y = a + bx$), select:
   9:LinReg(a+bx) LINR
   (b) For logarithmic regression ($y = a + b \ln x$), select
   0:LnReg LNR
   (c) For Exponential Regression ($y = ab^x$), select
   A:ExpReg EXPR
   (d) For Power Regression ($y = ax^b$), select
   B:PwrReg PWRR
   (e) For Quadratic Regression ($y = ax^2 + bx + c$), select
   6:QuadReg P2REG

(f) For Cubic Regression ($y = ax^3 + bx^2 + cx + d$), select
    7:CubicReg   P3REG
(g) For Quartic Regression ($y = ax^4 + bx^3 + dx^2 + ex + f$), select
    8:QuartReg   P4REG

On the TI-82, after making the selection, press ENTER.

The calculator gives values for constants in the equations and the correlation coefficient.

❖ Only (a), (b), (c), and (d) are available on the TI-81.

Procedure S8. To graph data and the related regression equation.

1. Make sure the viewing rectangle is set properly to graph the set of data and that the graph display has been cleared, and that all functions in the function table have been removed or turned off.
2. If the data has not already been entered into the calculator, it will need to be entered (Procedure S6) and the linear regression equation calculated (Procedure S7).
3. Store the regression equation for a function name and graph by:

On the TI-82:

(a) Press the key Y=
(b) Move the cursor after an unused function name.
(c) Press the key VARS
(d) From the menu, select 5:Statistics...
(e) Move the highlight in the top row to EQ
(f) Select 7:RegEQ
(g) Press the key GRAPH

On the TI-85:

(a) Press the key GRAPH
(b) From the menu, select y(x)=
(c) Move the cursor after an unused function name.
(d) Press the key VARS
(e) From the menu, select STAT (after pressing MORE twice).
(f) Move the mark on the left to RegEq
(g) Press ENTER
(h) Select Graph from the graph menu.

❖ On the TI-81, aftering pressing VARS, highlight LR in top row and select 4:RegEQ.

4. Define the plot and draw a scatter graph:

On the TI-82:

(a) Press the key STAT PLOT

(b) Select one of the three plots.

The screen will show the plot number selected, followed by several options. The highlighted options are active. (To make an option active, move the cursor to that option and press ENTER.)

(c) Highlight the option ON.

(d) Highlight the type of graph. The first symbol following the word Type is for a scatter plot.

(e) For Xlist, highlight the name of the x-values.

(f) For Ylist, make the name of the y-values active.

(g) For Mark, make active the symbol to be used in plotting data.

(h) To exit, press QUIT

(i) To see the graph, press GRAPH.

On the TI-85:

(a) Press the key STAT

(b) From the menu, select DRAW

(c) From the secondary menu, to draw a graph, select:
SCAT for a scatter graph.

❖ On the TI-81, press STAT, highlight DRAW, and select 2:Scatter and press ENTER.

Appendix B

Procedures for Casio 7700 Calculator

B.1 Basic Calculator Procedures

Procedure B1. To turn calculator on and off.

1. To turn the calculator on, press the key: AC
2. To turn the calculator off, you will need to first press the key: SHIFT then the key: AC

Procedure B2. To clear memory in the calculator.

1. Turn the calculator on.
2. Press the key: SHIFT
3. Press the key: CLR
   This gives the menu: Mcl Scl Arr Prg
   Select Mcl to clear all value memory
   Select Scl to clear statistical memory
   Select ARR to clear matrix memory
   Select PRG to clear all program memory
4. Press EXE to clear the respective memory.

Note: The Casio 7700 has a maximum of 4164 bytes of memory available. To see the remaining number of bytes, press SHIFT, then press the key Defm

Procedure B3. Adjusting the contrast of the screen.

To adjust the contrast of the screen:
Press the key: MODE then

1. To make the screen lighter:
   Press the left cursor key.
2. To make the screen darker:
   Press the right cursor key
   The contrast change takes place in a series of increments. Keep pressing the key to obtain the contrast you desire.
   When finished press: AC

Procedure B4. To Clear the Calculator Screen

To clear the calculator screen, press the key: AC

B.2 General Calculator Procedures

Procedure C1. To perform secondary actions or obtain alpha characters.

1. To perform secondary actions of keys:
   (The actions above and to the left of a key.)
   a) Press the key: SHIFT
   b) then, press the key for the desired action
      (Example: To find the square a number we would first enter the number, then press the key: SHIFT then press the key: √ Then press EXE)
2. To obtain alpha characters:
   (The characters above and to the right of a key.)
   (a) Press the key: ALPHA
   (b) then, press the key for the desired character.
      (Example: To display the letter B on the screen, press the key: ALPHA then press the key: log)
      Note: The alpha key may be locked down by first pressing the key: SHIFT then pressing the key: ALPHA

Procedure C2. Selection of menu items

The Casio 7700 list menus at the bottom of the screen when certain keys are pressed. To select a menu item, press the function key (F1, F2, F3, F4, F5, F6) just below the item.
(Example: To display the symbol μ on the screen, press the key: ENG SYM then press the key F2 to select μ.)
To exit a menu, press the key: PRE for the previous screen.

Procedure C3. To perform arithmetic operations.

1. To add, subtract, multiply or divide press the appropriate key: +, –, ×, or ÷. If parentheses are used to denote multiplication, it is not necessary to use the multiplication symbol. (The calculation 2(3 + 5) is the same as 2×(3 + 5).)
2. To find exponents use the key x^y.
   (Example: 2^4 is entered as $2x^y4$)
   For the special case of square we may use the key x^2.
   Note: On the Casio 7700 calculator, there is a key labeled (-) which may be used to enter a negative number.
3. After entering the expression press the key EXE to perform the calculation.

Procedure C4. To correct an error or change a character within an expression.

1. (a) To make changes in the current expression: Use the cursor keys to move the cursor to the position of the character to be changed.
   (b) If it is desired to make changes in the last expression entered, after pressing the ENTER key: Press the left or right cursor keys

to obtain a copy of the last expression.
Then, use the cursor keys to move the cursor to the character to be changed.

2. Characters may now be replaced, inserted, or deleted at the position of the cursor.
   (a) To replace a character at the cursor position, just press the new character.
   (b) To insert a character or characters in the position the cursor occupies, press the key: INS.
   Then, press keys for the desired character(s).
   (c) To delete a character in the position the cursor occupies, press the key: DEL

Procedure C5. Selection of special mathematical functions.

1. To raise a number to a power:
   Enter the number
   Press the key x^y
   Enter the power.
   Press EXE
2. To find the root of a number:
   (a) For cube root, press the key $\sqrt[3]{}$
   (b) For any root, enter the root index and press the key $\sqrt[x]{}$
   Enter the number
   Press EXE
3. To find the factorial of a positive integer:
   Enter the number
   Press the MATH key for the Math Menu
   Press F2 for the Probability Menu
   Press F1 to select x! from the menu
   Press ENTER.
4. To find the largest integer less than or equal to a given value (This is the greatest integer function and is indicated by square brackets []):
   Press the MATH key for the Math Menu
   Press F3 for the Numerical Menu
   Press F2 to select Int from the menu
   Enter the value
   Press ENTER.
5. To find use the number π, press the key for π on the keyboard.
6. To find the absolute value of a number.
   Press the MATH key for the Math Menu
   Press F3 for the Numerical Menu
   Press F1 to select Abs from the menu
   Enter the value
   Press ENTER.

Procedure C6. Changing the mode of your calculator.

Press the key: DISP

1. To fix the number of decimal places displayed:
   Press the key F1 to select Fix from the menu.
   Enter the number of decimal places desired. Press EXE
2. To display a fixed number of significant digits in engineering notation:
   Press the key F2 to select Sci from the menu.
   Enter the number of significant digits desired. Press EXE
3. To display a number in normal notation:
   Press the key F3 to select Nrm from the menu. Press EXE
4. To display numbers in engineering notation:
   Press the key F4 to select Eng from the menu. Press EXE

Note: Engineering notation uses M for millions, k for thousands, m for 1/1000 and μ for 1/1000000.

Procedures C7, C8, and C9 on lists.

Lists and operations with lists are not available on the Casio 7700.

Procedure C10. To store or recall a function.

A. To store a function.
   1. Enter the function.
   2. Press the key FMEM
   3. Press the key F1 for STO
   4. Enter a number from 1 to 6.
      (The Casio 7700 will store up to 6 functions.)

   The function will now be stored "permanently" in the calculator.
   To exit from FMEM, press the key PRE

B. To recall a stored function.
   1. Press the key FMEM
   2. Press the key F2 for RCL or press the key F3 for fn
   3. Enter a number from 1 to 6.
      (The Casio 7700 will store up to 6 functions.)

   The function or the function name will appear on the screen at the location of the cursor.
   To exit from FMEM, press the key PRE

Procedure C11. To evaluate a function.

After a function has been stored,

1. Store the x-value at which we wish to evaluate the function.
   (a) Enter the value.
   (b) Press the key →
   (c) Press the key X, θ, T

2. Evaluate the function.
   (a) Press the key: FMEM
   (b) Recall the a function:
       Press F2 to select RCL from the menu and enter the number of the function.
   (c) Press EXE to evaluate the function.

   The value of the function will appear at the right of the screen.

   Note: If an error message is obtained, press the key AC This likely indicates that the value of x is not in the domain of the real function.

To evaluate the same function for different x-values, repeat steps1and 2

Note: To see a list of all functions press F4 to select LIST from the FMEM menu.

Procedure C12. To store a constant for an alpha character.

1. Enter the value to be stored.
2. Press the key →
3. Enter the alpha character.
4. Press EXE.

(Example: 6.35→H)

Procedure C13. To obtain relation symbols (=, ≠, >, ≥, <, and ≤).

1. Press the key: PRGM
2. Press F2 to select REL from the menu.
3. Select the desired symbol from the menu by pressing a function key.

Procedure C14. To change from radian mode to degree mode and vice versa.

1. Press the key: DEG
2. Select from the menu:
   Press F1 for degrees
   Press F2 for radians
   Press F3 for gradients

Note: To determine if the calculator is in radian or degree mode, hold down the MDisp key. The third line indicates the angle mode.

Procedure C15. Conversion of coordinates.

Make sure calculator is in correct mode -- radians or degrees.

A. To change from polar to rectangular.
   1. Press the key Rec(
   2. Enter the value of r, a comma, then the value for θ.
   3. Press EXE

   The value shown is x. To see the value for y, press the key J, then press EXE.

   (Example: Rec(2.0, 45) gives 1.4 for x, 1.4 for y. [in degree mode])

B. To change from rectangular to polar.
   1. Press the key Pol(
   2. Enter the value of x, a comma, then the value for y.
   3. Press EXE

   The value shown is r. To see the value for θ, press the key J, then press EXE.
   (Example: Pol(1.0, 1.0) gives 1.4 for r, 45 for θ. [in degree mode])

Procedure C16. Finding the sum of vectors in polar form.

1. Find the x- and y-components of each vector (see Procedure C15).
2. Find the sum of the x- components and the sum of the y-components.
3. Find the magnitude and direction of the resultant vector (see Procedure C15).

Procedure C17. Conversion of complex numbers.

A. To change from polar to rectangular form:
   Think of the complex number as r and θ in polar form and use Procedure C15 to convert to rectangular form.

B. To change from rectangular to polar form:
   Think of the complex number as x and y in rectangular form and use Procedure C15 to convert to polar form.

Procedure C18. Operations on complex numbers.

1. If not already in rectangular form, change the complex number into rectangular form (Procedure C17).
2. Perform the operations on the rectangular forms of the complex numbers.
3. If the answer is to be in polar form, convert the result back to polar form (Procedure C17).

Procedure C19. To determine the numerical derivative from a function.

The Casio 7700 does not perform this operation.

Procedure C20. Function to calculate definite integral.

Press the key ∫dx . On the screen appears ∫(
The form of the function is ∫(y, a, b, n).
where y is the function which may be entered or recalled from memory, a and b are the lower and upper limits of integration, and n is an optional number from 1 to 9 and relates to the accuracy of the calculation. (The higher the value, the more accuracy.)

(Example: ∫(x^3, 1, 3) is used to evaluate $\int_1^3 x^3\,dx$.)

B.3 Graphing Procedures

Procedure G1. To graph functions.

A. To graph a single function.
   1. Press the key: GRAPH
   2. Enter the function to be graphed.
      (Example: to graph $y = x^2 - 4$, enter $x^2 - 4$ after Graph Y=)
   3. Press EXE

B. To graph more than one function.
   1. Press the key: Graph
   2. Enter the first function to be graphed after the equal sign.
   3. Place a colon after this function by pressing the key: PRGM then press F6 to select **:** from the menu.
   4. Repeat steps 1, 2, and 3 for each function to be graphed.

   Functions may also be stored and recalled from memory (Procedure C10).

(Example: To graph $y = 2x$ and $y = 2x-4$ the following should appear on the screen: Graph Y=2X:Graph Y=2X–4)

Notes:

1. Functions stored in memory will remain in the calculator's memory until changed or erased, or until the batteries run down or are removed.
2. While the graph is on the screen, pressing the key: G↔T will clear the graph off the screen. However, the graph is still in the calculator's memory and can be seen again by pressing the key: G↔T

Procedure G2. To change or erase a function.

1. To change a function:
   Recall the function (Procedure C10), make desired changes, and store the revised function for the same function number.
2. To erase a function:
   (a) Press the key: FMEM
   (b) Press the key: AC (clears screen)
   (c) Press F1 to select STO from the menu.
   (d) Enter the function number.
3. To exit the function menu or table, press PRE

Procedure G3. To obtain the standard viewing rectangle.

1. Press the key RANGE.
2. Press the key F1 to obtain INIT from the menu.

Procedure G4. To use the trace function and find points on a graph.

1. With a new graph on the screen press the key F1 for TRACE. This will cause the cursor to appear on the graph.

2. As the right and left cursor keys are pressed, the cursor moves along the graph and the coordinates of the cursor are given on the bottom of the screen. Pressing the key F5 while the coordinates are on the screen will give a more accurate value of the x-coordinate, pressing F5 a second time will give a more accurate value of the y-coordinate, pressing F5 a third time will return to the original coordinates,
3. If more than one graph is on the screen, pressing the up and down cursor keys will cause the cursor to jump between the graphs.
4. If the cursor is moved off the right or left of the screen the graph scrolls (moves) right or left to keep the cursor on the screen. The cursor will not go off the top or bottom of the screen.
5. Pressing the cursor keys without first pressing TRACE, will cause the viewing rectangle to move in that direction.
6. Pressing the F5 key while a graph is on the screen will cause CLS to be displayed. If EXE is then pressed, the graph will be erased.

Procedure G5. To see or change viewing rectangle.

1. Press the key: RANGE
2. For each of the quantities you wish to change enter the value for that quantity. To keep a value and not change it, either
   (a) Press the key: EXE or
   (b) Use the cursor key to move the cursor to a value to be changed and enter the new value.
3. To see the graph, after the new values have been entered, press the key: GRAPH
4. To exit from RANGE, continue to press EXE till you have passed through the entire list or press the key: RANGE (this may need to be done twice).
   The initial viewing rectangle on the Casio 7700 has the following values:

Xmin = –4.7	Ymin = –3.1	T,θmin = 0
Xmax = 4.7	Ymax = 3.1	T,θmax = 2π (or 360°)
Xscl = 1	Yscl = 1	ptch = 2π/100 (or 3.6°)

T,θmin, T,θmax and ptch are needed for parametric and polar graphing. For normal graphing they may be left unchanged.

Procedure G6. To obtain special viewing rectangles.

1. Press the key: RANGE
2. For the standard viewing rectangle:

   Xmin = –10, Xmax = 10, Xscl = 1,
   Ymin = –10, Ymax = 10, Yscl = 1

   Enter these values.
3. For the initial viewing rectangle:
   Press the key F1 to select INIT from the menu (see Procedure G3).

4. For the trigonometric viewing rectangle:
   Press the key: GRAPH, then press the key: sin and then press EXE.
   This will graph the y = sin x on the trigonometric viewing rectangle.
   Press CLS, then press EXE to clear the graph.
   The preset values for the trigonometric viewing rectangle will remain in the memory. Values for the trigonometric viewing rectangle are:

   Xmin = -2π (or $-360°$)   Xmax = 2π (or 360°)   Xscl = π
   Ymin = -1.6 Ymax = 1.6   Yscl = .5

Procedure G7. To graph functions on an interval.

1. To graph a function on the interval $x < a$ or on the interval $x \le a$ for some constant a:
   In the function table, enter the function f(x) followed by ,[k, a] where k is any value smaller than Xmin.
   (Example: To graph $y = x^2$ on the interval $x<2$, enter for the function: x^2, [−100, 2])
2. To graph a function on the interval $a < x < b$ or on the interval $a \le x \le b$ for some constants a and b:
   In the function table, enter the function f(x) followed by ,[a, b].
   (Example: To graph $y = x^2$on the interval $3 \le x \le 2$, enter: x^2,[−3, 2])
3. To graph a function on the interval $x > a$ or on the interval $x \ge a$ for some constant a:
   In the function table after the equal sign, enter the function f(x) followed by ,[a, k] where k is greater than Xmax.
   (Example: To graph $y = x - 5$ on the interval $x>2$, enter for the function: (x−5),[2, 100])

Note: Piecewise functions need to be graphed as separate functions on separate intervals.

Procedure G8. Changing graphing modes.

Press MODE, then press SHIFT, then:
1. Press the key 5 for plotted points to be connected.
   or
   Press the key 6 to plot individual points.
2. Press the key + for rectangular coordinate graphing.
3. Press the key − for polar coordinate graphing.
4. Press the key × for graphing parametric equations.
5. Press the key ÷ for graphing inequalities

Press PRE to exit.

Procedure G9. To zoom in or out on a section of a graph.

1. Graph a function(s).
2. With a graph on the screen, press the key ZOOM

3. Press TRACE and move the cursor near the desired point. After zooming this point will be near the center of the screen.
4. To zoom in: Press F3 to select xf from the menu or
   To zoom out: Press F4 to select x/f from the menu.
5. To continue zooming, repeat steps 3 and 4.

Procedure G10. To change zoom factors

1. Press the key: ZOOM
2. Press F2 to select FCT from the menu.
3. Change the factors as desired.
   The initial factors are 2 in both directions. For the magnification in the x-direction, change XFact and for magnification in the y-direction, change YFact.
4. To exit this screen press EXE

Procedure G11. Solving an equation in one variable.

1. Write the equation so that it is in the form f(x) = 0, let y = f(x), and graph the function. Use a viewing rectangle so that the x-intercept of interest appears on the screen.
2. Use Procedure G9 to zoom in on the x-intercept until the desired degree of precision is obtained. If the graph crosses the x-axis, it helps to use the trace function to determine points where the graph is positive and points where it is negative.

Procedure G12. Finding maximum and minimum points.

1. Graph the function and adjust the viewing rectangle so that the desired local maximum or local minimum point is on the screen.
2. Use Procedure G9 to zoom in on the point until the coordinates are determined to the desired precision. It helps to use the trace function to move the cursor along the curve to determine the maximum or minimum point.

Procedure G13. Find the value of a function at a given value of x.

1. Graph the function and adjust the viewing rectangle so that the desired point is on the screen.
2. Use Procedure G9 to zoom in on the point to the desired degree of precision. Use the trace function to move the cursor to the given value of x.

Procedure G14. To zoom in using a box.

1. Press the key: ZOOM
2. Press F1 to select Box from the menu.
3. Use the cursor keys to move the cursor to a location where one corner of box is to be and press EXE

4. Use the cursor keys to move the cursor to the location for the opposite corner of box and press EXE. As the cursor keys are moved a box will be drawn and when EXE is pressed, the area in the box is enlarged to fill the screen.

Procedure G15. Finding an intersection point of two graphs.

1. Enter and graph both functions at the same time. Adjust the viewing rectangle so that the intersection point appears on the screen.
2. Use Procedure G9 to zoom in on the intersection point to the desired degree of precision. Use the trace function to aid in determining the point.

Procedure G16. To graph two functions and their sum.

1. Store the first function into the function table for f_1. (Procedure G2)
2. Store the second function into the function table for f_2.
3. Start the graphing statement to graph these two functions. (On the screen: Graph f_1: Graph f_2:)
3. Press FMEM, then press GRAPH
4. Press F3 to select f_n from the menu, then press 1.
5. Press the key: +
6. Press F3 to select f_n from the menu, then press 2. (On the screen: Graph f_1: Graph f_2:Graph $Y=f_1+f_2$)
7. Press EXE

Procedure G17. To change graphing modes for regular functions, parametric equations, or polar equations.

1. Press the key MODE
2. Press the key SHIFT
3. For graphing in rectangular coordinates: Press + for REC
4. For graphing parametric equations: Press x for PARAM
5. For graphing in polar coordinates: Press – for POL
6. Press PRE to exit.

Procedure G18. To graph parametric equations.

1. Make sure calculator is in correct mode for parametric equations (Procedure G14) and graph is cleared.
2. Press the key: GRAPH
   This will give GRAPH(X,Y)=(
3. Enter the function for X, a comma, the function for Y and).
   When entering the functions, pressing the key X,θ,T will give the variable T.
4. Press EXE to graph the function.

(Example: If $x = t$ and $y = t^2$, GRAPH(X,Y)=(T,T^2))

Note: It may be necessary to adjust RANGE values. The functions may also be stored as f_1 and f_2 and graphed using GRAPH(X,Y)=(f_1,f_2)

Procedure G19. To change values for the parameter t.

With the calculator in parametric mode:

1. Press the key: RANGE
   The first screen gives the familiar values for x and y.
2. Press RANGE again. Now, we can enter values for Tmin, Tmax, and ptch. Ptch is the pitch and determines how far apart the points are plotted.
   The initial viewing rectangle values are: Tmin = 0, Tmax = 6.28.(2π) or 360°, and ptch = 0.0628..($2\pi/100$) or 3.6°.
3. To exit, press RANGE or EXE .

Procedure G20. To graph equations involving y^2.

1. Clear any previous graphs by pressing CLS while graph is on screen, then press EXE.
2. Solve the equation for y. There will be two functions: y = f(x) and y = g(x).
3. Store f(x) for f_1 and g(x) for f_2.
   Graph both functions (Procedure G3)

or

If g(x) = – f(x), we may do the following:
(a) Store f(x) for f_1 in the function table.
(b) Press the key: FMEM
(c) Press GRAPH
(d) Press F3 to select f_n from the menu.
(e) Press the key: 1
  (This will give: Graph Y=f_1)
(f) Press PRGM, then press F6 to select **:** from the menu.
(g) Press GRAPH, then press the key –
(h) Press F3 to select f_n from the menu.
(i) Press the key: 1
  (On the screen this appears as: Graph Y= f_1:Graph Y= $-f_1$)
(j) Press EXE to have the graph drawn.

Procedure G21. To obtain special Y-variables.

This does not pertain to the Casio 7700.

Procedure G22. To shade a region of a graph (graphing inequalities).

1. Clear the graphing screen (Press F5 for Cls while graph is on screen).
2. Change to inequality graphing mode (see Procedure G8).
3. When the graph key is pressed the following choices are given: Y>, Y<, Y≥, Y≤. Select the one desired.
4. Enter the function and press EXE. The graph will be drawn and the proper area shaded in. If two inequalities are graphed at the same time (separated by a colon :), the area satisfying both inequalities will be shaded.

5. A restricted domain may be specified for the graph by entering the minimum x and maximum x in square brackets after the function.
   Example: Graph Y>2x+3 [-2, 2]

Procedure G23. To graph in polar coordinates

1. Make sure the calculator is in the correct mode for polar graphing (Procedure G8) and graph screen is clear.
2. Press the key: GRAPH (This will give GRAPH r =)
3. Enter the function (as a function of θ).
   Pressing the key: X,θ,T will now give θ.
4. Press EXE.

Note: It may be necessary to adjust the RANGE values.

Procedure G24. To change values for θ.

With the calculator in polar mode:

1. Press the key: RANGE
   The first screen gives the familiar values for x and y.
2. Press RANGE again. Now, we can enter values for θmin, θmax, and ptch. Ptch is the pitch and determines how far apart the points are plotted.
   The initial viewing rectangle values are: θmin = 0, θmax = 6.28.(2π) or 360°, and ptch = 0.0628..($2\pi/100$) or 3.6°.
3. To exit, press RANGE or EXE .

Procedure G25. Drawing a tangent line at a point.

On the Casio 7700 it is necessary to find the equation of the tangent line and graph both the function and the tangent line on the same graph.

Procedure G26. Determining the value of the derivative from points on a graph.

This is not possible on the Casio 7700.

Procedure G27. Function to determine the area under a graph.

1. Press the key G-$\int$dx. The symbol $\int$ appears on the screen.
   The form is $\int$ y, lower, upper
2. Enter the function in place of y, the lower limit of integration for lower, and the upper limit of integration for upper.
   Example: $\int$ 2x+2, −1,3
3. Press EXE. The graph of the function is drawn and the area specified is shaded in. The value of the area is given at the bottom of the screen.

B.4 Programming Procedures

Procedure P1: To enter a new program.

1. Press the key MODE
2. To change to programming (WRT) mode, press the key: 2
   A list of programs will be seen. Additional programs can be seen by pressing the down cursor key.
3. Programs are selected by moving the cursor to the desired program or by pressing the number of the program. Select a program number to enter a new program (pick an empty slot).
4. Press EXE
5. On the first line on the screen enter the program name. It is limited to 12 characters. Press EXE
6. Enter a programming statement on each line. Press EXE after the statement. Or, several statements may be placed on one line by using a colon between statements. The colon is obtained from the PRGM menu. If it is not on the screen, try pressing the SHIFT key.
7. After the last statement in a program and on the same line, place the ◅ symbol.
8. Press MODE then the key: 1 to exit programming mode.

Procedure P2a. To edit program.

1. Press the key MODE
2. Change to programming mode, press the key: 2
3. Select a program to be edited and press EXE
   A listing of the program appears on the screen.
4. Use the cursor, insert, and delete keys to change (edit) the program.
5. Press the key MODE then the key: 1 to exit programming mode.

Procedure P2b. To execute or erase a program.

1. To execute a program:
   (a) You must be in RUN mode (Press MODE, then press 1)
   (b) Press the key: PRGM
   (c) Press F3 to select Prg from the menu.
   (For a listing of programs go to programming mode.)
   (d) Enter the program number and press EXE
2. To erase a program:
   (a) Press the key: MODE
   (b) Press the key: 3
   (c) Select a program and press the key: AC
   The selected program will be erased from memory.

Procedure P3. To input a value for a variable.

1. Press the key: PRGM (while in programming mode (WRT))
2. Press F4 to select ? from the PRGM menu.
3. Press the key: → (to store the value)

4. Enter an alpha character for the variable, press EXE
(Example: ?→C)
When the program is executed, the ? will cause a question mark to appear on the screen. Any number entered is stored for the given variable and when the variable is later used in the program that value is used in the calculation.

Procedure P4. To evaluate an expression and store that value to a variable.

1. Enter the expression to be evaluated.
2. Press the key: →
3. Enter the variable.

(Example: (9/5)C+32→F)

Procedure P5. To display quantities from a program.

Either the value for a variable or an alpha expression may be displayed.

1. The last variable stored or expression calculated before the ◅ symbol is displayed when the program is executed. To display more than one value, place the ◅ after each quantity to be displayed. If another value is to be displayed, the quantity -Disp- will appear on the screen. Press EXE to see additional values. The last value calculated is displayed if ◅ appears or not.
   (Example: (9/5)C+32→F◅ prints a value for F.)
2. To display characters as they appear, enter the characters inside of quotes. The quotes are obtained from a menu obtained by pressing the ALPHA key.
   (Example: "Enter C" will cause ENTER C to be displayed.)

Procedure P6. To add or delete line from a program.

1. Use Procedure P2 to bring a program to the screen so that it may be edited.
2. To add a line:
   (a) To insert a line before a given line move the cursor to the first character of the line and to insert a line after a given line move the character to the last character in the given line.
   (b) Press the key: INS then press EXE This will create a blank line either before or after the given line.
   (c) Make sure the cursor is in the desired line. An instruction may now be entered. When inserting a line before a given line, INS will need to be pressed again.
3. To delete a line
   (a) Move the cursor to the desired line
   (b) Press the key: DEL several times

Procedure P7. Finding the area under a graph.

1. Store the desired function for a function name.
2. Enter the following function on your calculator to find the area between two values B and C. (It is assumed that the function is stored under the function name f_1.)

```
AREA
"FIRST Z"
?→B
"SECOND Z"
?→C
"AREA IS"
∫(f1,B,C)
```

The symbol ∫(is obtained by pressing the key ∫dx.

Procedure P8. To display a graph from within a program.

1. Press the Graph key to obtain Graph Y=
2. Enter the function.

Procedure P9. Creating a loop.

While in programming edit mode (at a step in the program).

1. Press the key PRGM
2. Select JMP from the menu.
   From the next menu, select
   (a) The label (Lbl) command
   or
   (b) The goto (Gto) command.
   The *label* is to be an integer 1 to 9.
3. Select REL for the relation symbols.
   The form of a condition statement is: A rel B ⇒ C where A and B are compared by the rel(ation) symbols =, ≠, >, <, ≥, ≤. If A rel B is true, C is executed. If not, the next statement is executed.

The construction of a loop is:

... (program statements before the loop)
Lbl *label*
... (program statements within loop)
...
Condition statement
Goto *label*
... (program statements after loop)

Procedure P10. Drawing a line from within a program.

This does not pertain to the Casio 7700.

Procedure P11. To temporarily halt execution of a program. (Pause statement)

This does not pertain to the Casio 7700.

B.5 Matrix Procedures

Procedure M1. To enter or modify a matrix.

1. First the calculator must be put into Matrix mode. Do this by first pressing MODE, then press 0. The basic matrix menu appears at the bottom of the screen:

   A B + − × C

   Enter matrices for either A or B. C is used to store the results of calculations.
2. To enter a matrix press either F1 or F2 to select A or B from the menu. Pressing A, the following menu for matrix A appears:

   kA A^t |A| A^{-1} A↔B ◊

   If no matrix is in memory, the words *No existence* will appear.
3. Press F6 to select ◊ from the menu. This will give the menu:

   DIM ERS CLR ROW COL

   Press F1 to select DIM from the menu.
4. Enter the number of rows, press EXE,
   then enter number of columns, press EXE
   The matrix will now appear on the screen.
   The first element in the first row will be highlighted. This is the element to be entered.
5. Enter the first value in the first row and press EXE. The highlight will now move to the second element in the first row. Enter this value and press EXE. Repeat this process until all values have been entered.
6. To change a value, use the cursor keys to move the highlight to the value to be changed, enter the new value and press EXE.
7. When finished, press PRE to return to the previous screen and the menu as in step 1.

Note: Selecting CLR from the menu of Step 3 will make all elements of a matrix zero and press ERS will erase the matrix from memory. Selecting ROW or COL will allow you to delete, insert, or add a row or column to the matrix.

Procedure M2. To view a matrix.

1. To view a matrix, select A, B, or C from the basic matrix menu.
2. Press PRE to go back to previous screen.

Procedure M3. Addition, subtraction, scalar multiplication or multiplication of matrices.

The matrices to be added, subtracted, or multiplied must be stored for matrices A and B.

1. Addition matrices A and B:
   Select + from the basic matrix menu.
   The sum will appear on the screen and be stored as matrix C.
2. Subtraction matrix B from matrix A
   Select – from the basic matrix menu.
   The difference will appear on the screen and be stored as matrix C.
   Note: Matrix A cannot be subtracted from matrix B. To subtract A from B, first interchange A and B by selecting A↔B from the menu for matrix A (see Procedure M1, Step 2).
3. Multiplication of matrix by a scalar:
   To multiply matrix A by a scalar, enter the scalar, then select kA from the menu for matrix A. The scalar product will be stored for matrix C. Likewise, for matrix B.
4. Multiplication of matrix A by matrix B:
   Select X from the basic matrix menu.
   The product AB will appear on the screen and be stored as matrix C. To multiply B by A, first interchange matrix A with matrix B. (see note in Step 2)
5. Combinations of operations:
   To do combinations of operations, it is necessary to do one operation at a time with the result being stored as matrix C. The contents of matrix C may then be copied to either matrix A or matrix B by selecting C→A or C→B from the menu for matrix C. This may then be operated on by another matrix.
   (Example: To do 3A – 2B: Find 3A (Step 3), copy C to A (see Step 5), find 2B, copy C to B, then find A – B.)

Procedure M4. To find the determinant, inverse, and transpose of a matrix.

Select from the menu for the matrix. (See Procedure M1, Step 2)

1. To find the determinant of matrix A:
   Select |A| from the menu for matrix A.
2. To find the inverse of matrix A:
   Select A^{-1} from the menu for matrix A.
   The inverse will be stored as matrix C.
3. To find the transpose of matrix A:
   Select A^t from the menu for matrix A.
   The transpose will be stored as matrix C.

For matrix B, select similar items from the menu for matrix B.

Procedure M5. To store a matrix.

Matrices A and B may be interchanged by selecting A↔B from the menu for either A or B. There is no other way of storing a matrix.

Procedure M6. Row operations on a matrix.

On the Casio 7700 there is no provision for doing row operations on matrices.

B.6 Statistical Procedures

Procedure S1. To enter or change one variable statistical data.

1. First change the calculator to one variable statistical (SD) mode by pressing the MODE key, then pressing the key: ×
   The basic one variable statistic menu appears as

   DT EDIT ; CAL

   or as

   DT EDIT ; DEV Σ CAL

2. We will assume that the data entered is to be stored in memory. To have the data stored, press the MODE key, then press SHIFT, then press the key: 1 (If it is not to be stored, press 2.)
3. To enter data:
   After each data value is entered, select DT from the basic one variable statistic menu.
   DT enters the data value into memory.
   If it is desired to enter values with a frequency greater than one, enter the value, select ; from the basic statistic menu, enter the frequency, then select DT from the basic statistic menu. Repeat until all values have been entered.
   (Example: 3 ; 3 DT 4 ; 2 DT etc.)
5. To display the data that has been stored, select EDIT from the basic statistic menu. Press PRE to return to the previous screen.
6. To change data values that have been entered:
   (a) Select EDIT from the basic statistic menu.
   (b) Use the cursor keys to move the highlight to the value to be changed. (The highlighted value appears at the bottom right of the screen.)
   (c) Enter the new value and press EXE
   (d) When finished, press PRE

Note: When viewing data, only 5 lines of data appear at one time on the screen, but pressing the up or down cursor key will scroll the data on the screen.

Procedure S2. To graph single variable statistical data.

1. First make sure the calculator is in Draw mode by pressing MODE, then press SHIFT, then press the key: 3
2. Make sure the RANGE values are of the correct magnitude for the graph.
3. Adjust the memory for the number of values entered by: Press Defm, enter the number of bars to be drawn, press EXE
4. Erase any data stored in the calculator by: Press EDIT, then select ERS from the EDIT menu, then select YES when asked to erase all.
5. Clear the statistical memory by: Press CLR, select Scl from the clear menu, press EXE
6. Now, enter the values as in Procedure S1.
7. (a) To draw a histogram: Press GRAPH, then press EXE
   (b) To draw a frequency polygon (or xyLine): Press GRAPH, press SHIFT, select LIN from the menu, and press EXE.

Note: For the graph to be drawn correctly, the above steps must be followed.

Procedure S3. To clear the graphics screen.

With the graph on the screen, press F5 to select CLS, then press EXE.

Procedure S4. To obtain mean, standard deviation and other statistical information.

1. From the one variable statistical menu, select CAL. The following menu appears:

   DT EDIT ; DEV Σ PQR

2. Select DEV from this menu.
   (a) For the mean, press F1 to select x from the menu.
   (b) For the σ-standard deviation, press F2 to select $_x\sigma_n$ from the menu.
   (c) For the s-standard deviation, press F3 to select $_x\sigma_{n-1}$ from the menu.
   (d) For other information, press F4 to select ◊ from the menu, then from the resulting menu:
   Select Mod for the mode
   Select Med for the median
   Select Max for the maximum value
   Select Min for the minimum value
3. Select Σ from the menu of Step 1, then
   (a) For the sum of the squares of the x-values, select Σx^2.
   (b) For the sum of the x's, select Σx
   (c) For the total number of values entered, select n
4. Press PRE to return to the previous screen.

Procedure S5. To clear statistical data from memory.

1. To erase stored data:
   Select EDIT from the one variable statistics menu, then press F3 to select ERS from the EDIT menu.
2. To clear the statistical memory:
   Press CLR, then press F2 to select Scl from the clear menu.

Procedure S6. To enter or change two variable data.

1. First change the calculator to two variable statistical (REG) mode by pressing the MODE key, then pressing the key: ÷
   The basic two variable statistic menu appears as

   DT EDIT ; CAL

2. We will assume that the data entered is to be stored in memory. To have the data stored, press the MODE key, then press SHIFT, then press the key: 1 (If it is not to be stored, press 2.)
3. To enter data:
   Enter the first x-value, a comma, then the related y-value and then select DT from the basic two variable statistic menu. DT enters the data value into memory.
   Repeat the procedure for the second set of coordinates, then the third, etc. Repeat until all values have been entered.
   (Example: 12,49 DT 16, 68 DT etc.)
5. To display the data that has been stored, select EDIT from the basic statistic menu. Press PRE to return to the previous screen.
6. To change data values that have been entered:
   (a) Select EDIT from the basic statistic menu.
   (b) Use the cursor keys to move the highlight to the value to be changed. (The highlighted value appears at the bottom right of the screen.)
   (c) Enter the new value and press EXE
   (d) When finished, press PRE

Note: When viewing data, only 5 lines of data appear at one time on the screen, but pressing the up or down cursor key will scroll the data on the screen.

Procedure S7. To determine a regression equation.

1. Select the type of regression equation desired:
   Press MODE, then
   (a) Press 4 for linear regression ($y = A + Bx$)
   (b) Press 5 for logarithmic regression ($y = A + B \ln x$)
   (c) Press 6 for exponential regression ($y = A e^{bx}$)
   (d) Press 7 for power regression ($y = A x^B$)

2. Select CAL from the two variable statistic menu (see Procedure S6, Step 1).
   This gives the following menu:

   DT EDIT ; DEV Σ REG

3. Select REG from this menu and from the regression menu:
   (a) Select A for the value of a in the regression equation, press EXE.
   (b) Select B for the value of b in the regression equation, press EXE.
   (c) Select r for the correlation coefficient, press EXE.

Procedure S8. To graph data and the related regression equation.

1. Make sure the RANGE values are properly set for the data and the type of regression equation has been selected.
2. The calculator needs to be in graphing mode. Press MODE, then press SHIFT, then press 3.
   Note: To make a scatter graph, the calculator must be in graphing mode before entering data. The data points will be plotted as the data is entered.
3. Enter the data (after first erasing any previous data).
4. Press GRAPH, then press SHIFT, select LIN from the menu, then enter 1 and press EXE. (On the screen: GRAPH=LINE1)
   Press F5 to select CLS while the graph is on the screen to clear the graph.

Note: The calculator needs to be set to graph in rectangular coordinates. (see Procedure G8)

Answers

To Selected Problems

Exercise 1.1

1. 358.7
2. 237.9
3. 15.89
4. 8.271
5. –3.18
6. 138.0
7. 0.69
8. 321.3
9. 22
10. –1460
11. 3.15
12. 0.5913
13. 4.84
14. 50.0
15. 11.6
16. 262
17. 7.2
18. 7.09
19. 76.3; 78.0
20. 84, 85
21. 53
22. 23000
23. 285.5
24. 5830
25. 196, 216, 235, 255
26. 50.4, 59.0, 68.0, 77.9, 88.2
27. 0.0, 20.0, 37.0, 37.78, 68.61
28. (a) 14.7 (b) 92.20
29. 129
30. 0.000465
31. 3.35
32. 16.27
33. 192
34. 3.56
35. 175.9
36. 2120
37. 479,001,600
38. 144
39. 4
40. 2002
41. 5040
42. $4.0329... \times 10^{26}$
43. 0.66
44. 27
45. 2
46. –2
47. 4
48. –2
49. {110, 68,105, 68}, {–42, 22 19, –22}, {2584, 1035, 2666, 1035}, {.45, 1.9, 1.4, .51}
50. {66.7, 56.8, 143.0, 92.1, 111.0}, {–21.9, 11.4, –32.0, –21.3, –66.9} { 992, 774, 4870, 2010, 1970}, {.506, 1.50, .635, .624, .248}
51. {38, 47, 130, 100, 210}, {1.8, 2.3, 6.5, 5.1, 10}, {–5, –2, 25, 16, 51}
52. {504, 775, 2010, 1750}, {10.0, 15.4, 39.9, 34.8}, {46.0, 58.1, 113.2, 101.8}
53. {506, 135, 1060, 4100}, {4.74, 3.41, 5.70, 8.00}

54. {11400, 1560, 34300. 262000}, {2.82, 2.26, 3.19, 4.00}
55. {68.0, 32.0, –40.0, 98.6, 212}
56. {283.50, 416.50, 23.75, 198.90}
57. –0.46
58. –.698
59. 10.7
60. 19.2
61. (5.34, –2.68)
62. (–7.87, 2.08)

Exercise 2.1

1. 6
3. 18.18, undefined
5. undefined, 0.75, –0.54, 0, 0.79
7. 2, not a real value, 3.32, not a real value, 0
9. 3, 5.48, 3, 2.24, 3.16
11. 0, 0.10, –6.1, 2.2, 3
13. 1, 0, –4, 0, –1
15. (a) 9.7, 5.4, 0.29 (b) 7.7, 3.4, –1.7
17. (a) 16.0, 17.2, 18.4, 19.6
    (b) 18.0, 19.4, 20.8, 22.2
19. 51.5 ft., 65.5 ft., 13.0 ft.
21. $w = 5874 - 2.17t$, 5841 t, 5828 t, 5820 t
23. $C = 23250 + 5.14l$, \$23584, \$24084, \$25178

Exercise 2.2

	(a)	(b)	(c)
1.	1.8	–6.3	none
3.	4.0	9.7	none
5.	2.6, –2.6	–7.0	(0.0, –7.0) [min]
7.	none	5.0	(1.5, 2.8) [min]
9.	0.0, 2.6, –2.6	0.0	(1.5, –7.1) [min] (–1.5, 7.1) [max]
11.	none	none	none
13.	3.5	1.4	none
15.	2.0	1.4	none

17. Parallel line are obtained. The lines have y-intercepts of 2, 4, and 6 respectively.
19. (a) straight line
    (b) line becomes steeper, increasing from left to right
    (c) line becomes steeper, decreasing from left to right
    (d) gives horizontal line
    (e) line moves vertically
    (f) line passes through origin
21. 5.4, 2.1

	x-int.	y-int.	f(-1)	f(1)
23.	none	c	c	c
25.	0	0	1	1
27.	0	0	-1	1
29.	none	none	-1	1
31.	$0 \le x < 1$	0	-1	1

Exercise 2.3

1. (a) −1.6 (b) −6.0 (c) (2.5, −12) [min]
   (d) D: all x R: $y \ge -12.25$ (e) dec: $x < 2.50$ inc: $x > 2.50$
3. (a) −16, 16 (b) 250 (c) (0.0, 250) [max]
   (d) D: all x R: $y \le 252$ (e) dec: $x > 0$ inc: $x < 0$
5. (a) −8.1, 8.1 (b) 0.0 (c) (4.7, −210) [min], (−4.7, 210) [max]
   (d) D: all x R: all y (e) dec: $-4.7 < x < 4.7$ inc: $x < -4.7$ and $x > 4.7$
7. (a) 0.75 (b) −1.0 (c) none
   (d) D: all x, $x \ne 3$ R: all y, $y \ne -4$ (e) inc: for all x, $x \ne 3$
9. (a) −4 (b) −4 (c) none
   (d) D: all x, $x \ne 4$ R: all y, $y \ne -8$ (e) dec: for all x, $x \ne 4$
11. (a) −3.9, 3.9 (b) 3.9 (c) (0.0, 3.9) [max]
   (d) D: $-3.9 \le x \le 3.9$ R: $0 \le y \le 3.9$ (e) dec: $0 < x < 3.9$ inc: $-3.9 < x < 0$
13. (a) −6.1, 15 (b) 9.6 (c) (4.5, 10.6) [max]
   (d) D: $-6.1 \le x \le 15$ R: $0 \le y \le 10.6$ (e) dec: $4.5 < x < 15$ inc: $-6.1 < x < 4.5$
15. (a) none (b) none (c) (1.7, 3.5) [min], (−1.7, −3.5) [max]
   (d) D: all x, $x \ne 0$ R: $y < -3.5$, $y > 3.5$
   (e) dec: $-1.7 < x < 0$, $0 < x < 1.7$ inc: $x < -1.7$, $x > 1.7$
17. (a) 3.1 (b) 7.5 (c) (3.1, 0.0) [min]
   (d) D: all x R: $y \ge 0$ (e) dec: $x < 3.1$ inc: $x > 3.1$
19. (a) −2.0 (b) 2.0 (c) none (d) D: all x, R: all y
   (e) inc: $x < 2$, $x > 2$
21. 18.37 L/h, 1678 r/min
23. (a) 606 feet, 2360 feet (b) 51 miles per hour, 24 miles per hour
25. (a) D: all x R: all y (b) inc: all x (c) −1, 0, 1
27. (a) D: $x \ge 0$ R: $x \ge 0$ (b) inc: $x > 0$ (c) not a real value, 0, 1
33. (a) D: all x R: $y \ge 0$ (b) dec: $x < 0$ inc: $x > 0$ (c) 1, 0, 1
31. (a) circle
   (b) both have x-intercepts of −5 and 5, for y_1: y-intercept is 5, for y_2: y-intercept is −5 (c) for y_1 D: $-5 \le x \le 5$ R: $0 \le x \le 5$ for y_2 D: $-5 \le x \le 5$ R: $-5 \le x \le 0$

Exercise 2.4

	(a)	(b)
1.	5.42, –4.42	(0.500, –24.3) [min]
3.	0.801	none
5.	1.67	none
7.	0.857	

9. 5.85, -0.854
11. 2.46
13. 4.84 years, \$7500, \$5100
15. 4.39 volts, v > 4.39, 0.0172, 0.0395
17. $V = 140x - 48x^2 + 4x^3$; 1.92 in

Exercise 2.5

1.
```
Disp "ENTER F"
Input F
5(F–32)/9→C
Disp "DEGREES C"
Disp F
```
Ans.: 100, 38, 37.0, –18, –40

3.
```
Disp "ENTER X"
Input X
X^4–X^3+X–2→Y
Disp "F(X) = "
Disp Y
```
Ans.: –2, –1, 24., 67.4, 38.5

5.
```
Disp "ENTER R"
Input R
Disp "ENTER H"
Input H
πR²H→V
Disp "VOL. IS"
Disp V
```
Ans.: 1,220,000, 1,150,000

7.
```
Disp "ENTER P"
Input P
Disp "ENTER T"
Input T
Disp "ENTER N"
Input N
Disp "ENTER R"
Input R
P(1+R/100/N)^NT-P_I
Disp "INTEREST IS"
Disp I
```
Ans.: \$376.13, \$377.69, \$378.46

Exercise 3.1

1. .399
3. .13
5. 1.25
7. .916
9. 46.4
11. 3.3814
13. 57.94°
15. 88.6°
17. 57.69°
19. not possible
21. 33.0°
23. 7.055°
25. (a) $\sin 23.5° = \cos 66.5° = .399$, $\cos 23.5° = \sin 66.5° = .917$
    (b) $\sin 17.45° = \cos 72.55° = .2999$ $\cos 17.45° = \sin 72.55° = .9540$
    (c) $\sin 77.8° = \cos 12.2° = .977$ $\cos 77.8° = \sin 12.2° = .211$
    Conclusion: $\sin A = \cos(90° - A)$ or $\cos A = \sin(90° - A)$
27. (a) 1 (b) 1 (c) 1
    Conclusion: for any angle A, $(\sin A)^2 + (\cos A)^2 = 1$
29. (a) 1 (b) 1 (c) 1
    Conclusion: for any angle A, $(\csc A)^2 - (\cot A)^2 = 1$
31. $x = 35.48°$
33. $x = 33.4°$
35. $x = 1.33$
37. $x = 24.36$
39. $x = 2.070$
41. $B = 38.1°$
43. $A = 72.9°$
45. 370 cm^2
47. $B = 66.59°$, $b = 58.32$, $c = 63.55$
49. $A = 44.2°$, $b = 1.61$, $a = 1.57$
51. $b = 183.4$, $A = 41.72°$, $B = 48.28°$
53. $c = .9942$, $A = 8.89°$, $B = 81.11°$
55. 9.46°, 7.13°, 5.71°, 4.76°, 4.09°
57. 38.0 ft, 33.8 ft, 30.4 ft, 27.6 ft, 25.4 ft

Exercise 4.1

1 $x = 1.19$, $y = -0.875$
3. $x = 11.8$, $y = 3.82$
5. Same line, dependent system
7. $x = 21.5$, $y = 27.3$
9. No solution, inconsistent system
11. Type A 55.0 cents, type B 76.8 cents
13. 4424 L of first grade, 7801 L of second grade
15. (a) No profit, loss of $615.50 (b) Profit of $603.50
    (c) (92.5, 3305), Profit is zero.

Exercise 5.1

1. (a) max: 1; 2; 3 min: -1; -2; -3 (b) 1; 2; 3 (c) changes the amplitude
3. (a) max; 2; 3; 2; 3 min: -2; -3; -2; -3 (b) 2; 3; 2; 3
   (c) reflects the graph about the x-axis

5. (a) 1.57 ($\frac{\pi}{2}$), 4.71 ($\frac{3\pi}{2}$);

   .785 ($\frac{\pi}{4}$), 2.36 ($\frac{3\pi}{4}$), 3.93 ($\frac{5\pi}{4}$), 5.50 ($\frac{7\pi}{4}$);

   .524 ($\frac{\pi}{6}$), 1.57 ($\frac{\pi}{2}$), 2.62 ($\frac{5\pi}{6}$), 3.67 ($\frac{7\pi}{6}$), 4.71 ($\frac{3\pi}{2}$), 5.76 ($\frac{11\pi}{6}$)

   (b) 4.0, 6.28 (2π); 4, 3.14 (π); 2, 2.09 ($\frac{2\pi}{3}$)

   (c) changes the period of the graph
7. (a) Period appears to be π. (b) 0.04π or 0.126 (c) Xmin = 0, Xmax = .251
9. Xmin = 0, Xmax = 0.502, Ymin = –5, Ymax = 5;
   x-intercepts are 0.0, 0.13, 0.25, 0.38, 0.50
11. Xmin = 0, Xmax = 167, Ymin = –0.75, Ymax = 0.75;
    x-intercepts are 20.8, 62.5, 104.2, 145.8
13. Xmin = 0, Xmax = 0.021, Ymin = –20, Ymax = 20; x-intercept is 0.011
15. (a) 15.0, –70.5 (b) 0.0, 0.0083, 0.0167 (c) 0.0042, 0.0208

Exercise 5.2

1. (a) 2, 6.28 (2π); 2, 6.28 (2π); 2, 6.28 (2π)

   (b) 3.14, 2.36, 1.57

   displacement first to second is –0.78 to left ($-\frac{\pi}{4}$)

   displacement first to third is – 1.57 ($\frac{\pi}{2}$)

   (c) shifts graph horizontally by –C units

   (d) amplitude has no effect
3. (a) 2, 3.14 (π); 2, 3.14 (π); 2, 3.14 (π)

   (b) 0.0, 0.78, 0.39

   displacement first to second is +0.78 ($\frac{\pi}{4}$)

   displacement first to third is +0.6 ($\frac{\pi}{8}$)

   (c) shifts graph horizontally by $-\frac{C}{B}$ units
5. (a) (1) 5 (2) 4π (3) 0

   (b) (1) 3 (2) 0.5 (3) +.25

   (c) (1) 2.5 (2) 2π (3) $+\frac{\pi}{4}$

   (d) (1) 3.5 (2) π (3) 0.25

7. (a) (1) 0.00, 1.00, 2.00, 3.00 (2) max: (0.50, 2.00); min: (1.50, −2.00)
   (b) (1) 1.28, 2.59 (2) max: (1.94, 3.50); min: (0.63, −3.50)
9. (a) 38.03, −43.13 (b) 0.88, 1.54; 2.79, 3.63
11. (a) 1400 N, 2000 S, 4500 S (b) at 29 min, 240 min, 280 min, 490 min
13. (a) 0.0069 s and 0.0153 s (b) 0.0028 s
    (c) 0.0194 s, difference = 0.0167 = period of graph
    (d) 0.0057 s, 0.0165 s, 0.0224 s (e) −100.5 v.
15. (a) 2.42 cm, −0.095 cm (b) 0.0342 s, 0.0440 s, 0.0787 s, 0.0885 s
17. (a) Y_2 is a reflection of Y_1 about the x-axis
    (b) same graph - functions are equivalent
    (c) same graph - functions are equivalent
    (d) same graph - functions are equivalent
    (e) graphs differ only by amplitude, x-intercepts same

Exercise 5.3

1.

	x-intercepts	local minimum	local maximum
(a) y = sin x	−6.2, −3.1 0, 3.1, 6.2 no asymptotes	(−1.6, −1.0) (4.7, −1.0)	(1.6, 1.00) (−4.7, 1.0)
y = csc x	none asymptotes at x = −3.1, x = 0.0, x = 3.1	(1.6, 1.0) (−4.7, 1.0)	(−1.6, −1.0) (4.7, −1.0)
(b) y = tan x	−3.1, 0.0, 3.1 asymptotes at x = −4.7, x = −1.6, x = 1.6, x = 4.7	none	none
y = cot x	−4.7, −1.6, 1.6, 4.7 asymptotes at x = −3.1, x = 0, x = 3.1	none	none
(c) y = cos x	−4.7, −1.6, 1.6, 4.7 no asymptotes	(−3.1, −1.0) (3.1, −1.0)	(0.0, 1.0) (6.3, 1.0) (-6.3, 1.0)
y = sec x	none asymptotes at x = −4.7, x = −1.6, x = 1.6, x = 4.7	(−6.3, 1.0) (0.0, 1.0) (6.3, 1.0)	(−3.1, −1.0) (3.1, −1.0)

3. As A becomes larger, the graph stretches. Zeros and asymptotes stay same.
5. The graphs have different periods. As B doubles, the period is cut in half.
7. As C changes, the graph is displaced by −C units.
9. The negative sign reflects the graph about the x-axis.
11. (a) 3.57 cm, 3.72 cm (b) min b = 3.01 cm when A = 1.57 radians.
    (c) As A gets closer to π, b becomes very large.

13. max: (0.00, 3.00) and (6.28, 3.00), min: (3.14, 1.00), x-int: none
15. max: (0.00, 1.00) and (3.17, 1.20), min: (1.56, −0.95), x-int: 0.79, 2.30
17. max: (0.37, 1.63), (2.77, 1.63), (4.71, 0.00);
    min: (1.57, −2.00), (3.91, −0.71), (5.52, −0.71)
    x-int: 0.94, 2.20, 3.46, 4.71, 5.97
19. max: (0.65, 5.71), (4.31, 3.59); min: (1.97, −3.59), (5.63, −5.71)
    x-int: 0.00, 1.36, 3.14, 4.92, 6.28
21. max: (0.12, 3.58), (2.20, 0.12), (4.00, 2.88);
    min: (1.28, −1.94), (2.91, −0.90), (5.20, −3.95)
    x-int: 0.79, 2.04, 2.35, 3.30, 4.56, 5.81
23. (a) (2.02, 2.34) (b) 0.55, 1.50, 1.57, 1.64
25. (a) max is 3.01 when t = 2.20 (b) highest – lowest = 6.71

Exercise 5.4

1. (a) Gives line from (−14, 9) to (4,3) (b) no x-int, y-int = 4.33
3. (a) Part of parabola $y = 9x^2$ from (−15, 25) to (15, 25)
   (b) x-int: 0.00, y-int: 0.00
5. (a) Part of hyperbola y = 1/x
   (b) No intercepts, end points (−32, −0.50), (32, .50)
7. Graph is ellipse. x-int: −3.0, 3.0 y-int: −5.0, 5.0
9. (a) Graph is line from (−3, −5) to (3, 5)
   (b) Graph is like figure eight along y-axis with two loops.
   (c) Graph is figure along y-axis with four loops.
   Do not get closed loops if A is odd.
11. (a) Graph is a circle of radius 5 with center at (0,0).
    (b) Graph is an ellipse at a 45° angle.
    (c) Graph is a line from (−5, −5) to (5, 5)
    As C gets closer to π/2, graphs approaches a straight line at 45° angle.
    When C = π, graph is a circle again.

Exercise 6,1

1. [5.91, 120.9]
3. [−0.118, −0.129]
5. [−12146, 1534.5]
7. [0.00218, 1.25]
9. [30.1 ∠56.8°]
11. [1.485 ∠186.8°]
13. [1279 ∠5.71°]
15. [158 ∠164°]
17. [29.9 ∠62.0°]
19. [63.11 ∠170.49°]
21. [9.58 ∠156.5°]
23. [1223 ∠154.43°]
25. horz: 18.0, vert: 16.4
27. 115.7 east, 88.27 north
29. [119 ∠17.8°]

Exercise 6.2

1. $4.5 \angle 39°$
3. $199.4 \angle 128.3°$
5. $0.139\angle 206°$
7. $-437 + 714j$
9. $-1.111 - 1.225j$
11. $-80.0 + 36.3j$
13. $25.6 - 7.60j$
15. $-9654 + 10037j$
17. $0.828 - 0.878j$
19. $27.6 \angle 111.5°$
21. $305 \angle 50.2°$
23. $9840\angle 122.8°$
25. (a) $8 \angle 0°, 8\angle 0°$ (b) yes, $-1 + 1.732j, -1 - 1.732j$
27. $300 + 26.0j$
29. $-32.6 + 533j$
31. $45.4 \angle 44.7° = 32.3 + 31.9j$

Exercise 7.1

1. (a) For larger base number, graphs increase slower for $x < 0$ and faster for $x > 0$.
   (b) Inc: all x
   (c) Asymptote is $y = 0$
   (d) no x-intercept, y-intercept is 1
   (e) domain: all real nos.; range: $y > 0$
3. (a) For larger base number, graphs increase faster for $0 < x < 1$ and slower for $x > 1$.
   (b) Inc: all x
   (c) Asymptote is $x = 0$
   (d) x-intercept at $x = 1$; no y-intercept
   (e) domain: $x > 0$; range: all real nos.
5. (a)-(b) $y = 4^x$ continuously increases and $y = 4^{-x}$ decreases;
   (c) In each case asymptote is $y = 0$; They are same.
   (d) In each case y-intercept is $y = 1$; no x-intercept.
   (e) In each case domain: all real numbers; range: $y > 0$
   (f) Yes, reflection about the y-axis.
9. Graphs are reflection about $y = x$. Functions are inverse functions.
11. Graphs are reflection about $y = x$. Functions are inverse functions.
13. $V = 1225(1.0225)^{2t}$ (a) \$1635.91, \$1828.42 (b) 15.6 years
15. (a) 54.3 m/s, 73.8 m/s (b) Asymptote is $y = 95$, limiting velocity is 95 m/s

Exercise 7.2

1. 0.803
3. 4.21093
5. 0.548
7. -2.7881
9. 2.6233
11. 8.7147
13. 133000
15. 1.3330
17. 4.618
19. 6.14×10^{-6}
21. 4.53
23. 855

25. a) 0.3713
    b) 1.3713
    c) 2.3713
    d) 3.3713
    e) 9.3713
    All the decimal portions (mantissas) are the same.
27. 3.17
29. 1.8754
31. 3.5
33. 0.234
35. (a) 15.4 decibels (b) 2.57 W
37. (a) 17.4 yrs (b) 11.6 yrs (c) 8.75 yrs

Exercise 8.1

1. x = 2.79, y = −0.196; x = −3.46, y = 3.97
3. x = 2.75, y = 4.17; x = −2.75, y = 4.17; x = 1.78, y = −4.67; x = −1.78, y = −4.67
5. x = 5.25, y = 3.15; x = −5.25, y = 3.15; x = 3.01, y = −1.48; x = −3.01, y = −1.48
7. x = 4. 02, y = −14.1; x = −.84, y = 15.2
9. x = 1.84, y = 4.36; x = −1.84, y = −4.36
11. x = 1.41, y = 0.987; x = −0.637, y = −0.595
13. x = 1.41, y = 0.148; x = 0.0386, y = −1.41
15. x = 5.22
17. The helicopter is 1.694 miles east and 5.083 miles north of the tower.
19. (31.8, 7.52) and (23.8, 13.2); 32.8 ft and 57.6 ft.

Exercise 9.1

1. A is 3×3; B is 3×1; C is 3×3; D is 1×4; E is 1×4
3. −1 is a_{23}, c_{13}, and d_{13}
5. A and C; D and E
7. $\begin{bmatrix} 7 & -5 & -5 \\ 6 & 6 & -6 \\ 8 & -7 & 14 \end{bmatrix}$
9. not possible, not same size
11. $\begin{bmatrix} -8 & 22 & -14 \\ 6 & 0 & 6 \\ 14 & 2 & 14 \end{bmatrix}$
13. $\begin{bmatrix} -25 & 7 & 31 \\ -30 & -26 & 22 \\ -44 & 29 & -70 \end{bmatrix}$
15. $\begin{bmatrix} -14 \\ 18 \\ 62 \end{bmatrix}$
17. $\begin{bmatrix} 0 & 21 & -39 \\ 21 & -8 & -20 \\ 45 & -78 & 54 \end{bmatrix}$

19. $\begin{bmatrix} -165 \\ -132 \\ 66 \end{bmatrix}$

21. $\begin{bmatrix} -18 \\ 11 \\ 54 \end{bmatrix}$

Exercise 9.2

	transpose	determinant	inverse matrix
1.	$\begin{bmatrix} 2 & -3 \\ -5 & 4 \end{bmatrix}$	-7	$\begin{bmatrix} -.571 & -.714 \\ -.429 & -.286 \end{bmatrix}$
3.	$\begin{bmatrix} 3 & -4 & 5 \\ 2 & 0 & 3 \\ -1 & 3 & -2 \end{bmatrix}$	-1	$\begin{bmatrix} 9.000 & -1.000 & -6.000 \\ -7.000 & 1.000 & 5.000 \\ 12.000 & -1.000 & -8.000 \end{bmatrix}$
5.	$\begin{bmatrix} 1 & -5 & 7 \\ -2 & 3 & -7 \\ -4 & 1 & -9 \end{bmatrix}$	0	No inverse
7.	$\begin{bmatrix} .03 & -.04 & .11 \\ .12 & .45 & .95 \\ .10 & .32 & .04 \end{bmatrix}$	$-.013$	$\begin{bmatrix} 22.147 & -6.985 & .511 \\ -2.850 & .759 & 1.053 \\ 6.776 & 1.185 & -1.417 \end{bmatrix}$

	determinant	inverse	solution
9.	-13	$\begin{bmatrix} .230 & .154 \\ .154 & -.231 \end{bmatrix}$	$x = 2.231$, $y = -.846$
11.	0	none	no unique solution
13.	-114	$\begin{bmatrix} -.237 & -.447 & .026 \\ .053 & .211 & .105 \\ .272 & .254 & .044 \end{bmatrix}$	$x = -6.789$, $y = 3.842$, $z = 5.351$
15.	6	$\begin{bmatrix} 1.500 & -1.333 & .833 \\ 1.500 & -1.667 & 1.167 \\ .500 & -.667 & .167 \end{bmatrix}$	$x = -10$, $y = -9$, $z = -5$

17. $-.00905$ $\quad \begin{bmatrix} 4.111 & 1.547 & -1.149 \\ 5.438 & -2.255 & -6.897 \\ 12.467 & -7.405 & -12.356 \end{bmatrix}$ $\quad x = 1024.3$, $y = -392.6$, $z = -402.1$

19. -11.97 $\quad \begin{bmatrix} .253 & 1.553 & .062 \\ -.165 & 1.036 & .042 \\ -.0025 & 2.757 & -.050 \end{bmatrix}$ $\quad x = .0627$, $y = .0418$, $z = -.0501$

21. $\begin{bmatrix} 0.793 & 0.609 & 0 \\ -.609 & .793 & 0 \\ 0 & 0 & 1 \end{bmatrix}$

23. x = 89.1 g, y = 78.1 g, z = 32.8 g

Exercise 9.3

1. x = 5.23, y = 3.27
3. No solution.
5. x = –4.5, y = 5.7, z = 7.8
7. Infinite Solutions, x = 0.333z + 5.227, y = 1.333z + 0.867
9. x = 11.5, y = –3.7, z = 7.2
11. x = 14.5, y = 0.0, z = 1.5
13. 312 parts/h, 228 parts/h, 132 parts/h

Exercise 10.1

11. Corner points: (0, 0), (9, 0), (0, 10.4)
13. Corner points: (3.92, 0), (0, 2.35)
15. Corner points: (0, 0), (3.80, 0), (1.40, 3.00), (0, 3.00)
17. Corner points: (0, 0), (7.14, 0), (3.68, 6.06), (0, 6.80)
19. Corner points: (15.0, 0), (65.0, 0), (9.22, 11.2), (13.7, 0.636)
21. Corner points: (2.36, 5.57), (–1.69, 2.86)
23. Corner points: (0.785, 0.707), (3.92, 0.707)
25. $x + y \leq 185$, $x \geq 0$, $y \geq 0$; Corner points: (0, 0), (185, 0), (0, 185)
27. $125x + 158y \leq 47500$, $x \geq 100$, $y \geq 0$;
    Corner points: (100, 0), (380, 0), (100, 35000)
29. $x \geq 0$, $x \leq 322$, $175 \leq y \leq 475$;
    Corner points: (0, 175), (322, 175), (322,475), (0,475)

Exercise 11.1

1. (a) circle, radius = 2 (b) 4
   (c) (0,0), (4,0)
3. (a) 4-petal rose (b) 6
   (c) (6, 0), (6, 1.57), (6, 3.14), (6, 4.71), (0, 0)
5. (a) A cardioid (b) 10
   (c) (6, 0), (10, 1.57), (6, 3.14), (2, 4.71)
7. (a) Ellipse, F(0, 0) (b) 4.71
   (c) (0.2, 0), (0.17, 1.57), (0.2, 3.14), (.25, 4.71)
9. (a) Spiral (b) infinite
   (c) (0,0), (9.87, 1.57), (39.5, 3.14), (88.8, 4.71)
11. Circles with center on polar axis and passing through pole. As A changes, radius of circle changes. Radius = $\frac{A}{2}$.
13. Roses of 4 petals, 8 petals, and 12 petals. B determines the number of petals the graph has. If B is even, number of petals = B. Doesn't hold for B odd.
15. Cardioid, vertex at $(2, \frac{\pi}{2})$; Cardioid, vertex at (0, 0); limacon with an inner loop. If $|A| > |B|$, then cardioid with vertex not at origin; If $|A| = |B|$, then cardioid with vertex at origin; If $|A| < |B|$, then cardioid with inner loop.
17. Ellipse, parabola, hyperbola with vertices at (0, 0). If $|B| > |C|$, then ellipse; If $|B| = |C|$, then parabola; If $|B| < |C|$, then hyperbola.
19. (0, 0), (3.54, 0.79)
21. Spiral, (0.49, 1.57), (1.97, 3.14), (4.44, 4.71)
23. Min. dist. = 3.5, max. dist. = 13.9
25. distance along x-axis = 6.50 m, angle = 57.8°

Exercise 12.1

1. (a)

x	5	6	7	8	9	10
f	1	1	3	3	1	3

   (e) 7.92 (f) 8 (g) 7, 8, 10 (h) 1.62

3. (a)

x	75	80	85	90	95	100
f	1	2	2	3	1	2

   (e) 88.2 (f) 90 (g) 90 (h) 8.15

5. (a)

x	75	80	85	90	95	100
f	3	1	2	3	1	2

   (e) 86.7 (f) 87.5 (g) 75, 90 (h) 9.13

7. (a)

no. of instructions	18	19	20	21	22	23	24	25
frequency	2	3	0	2	3	1	3	1

   (e) 21.4 (f) 22 (g) 19, 22, 24 (h) 2.35
9. (a) 22.6 (b) 0.84 (c) 45 (86.5% of values)
11. (a) $383.90 (b) 65.3 (c) 10 (71.4% of values)

13. mean = 3.32", median = 3.31", mode = 3.32", s = 0.027", 90 (56.6% of values)
15. mean = 1201 hr., median = 1202.5 hr., mode = 1205 hr., s = 20.1 hr., 870 (64.3% of values)

Exercise 12.2

1. $y = 2.97 + 0.251\,x$, $r = 0.968$, good fit
3. $p = 650.5 - 0.2056x$, $r = -.999$, very good fit
5. $y = 3.069 + 1.970 \ln x$, $r = 0.996$, good fit, $y = 7.61$
7. $y = 10.15\,(2.13)^x$, $r = 0.99988$, very good fit, $y = 209$
9. $y = 1442.5x^{-1.01}$, $r = -0.99946$, very good fit, $P = \$2.72$
11. linear, $R = 0.8471 + 0.008240\,T$
    (a) 0.929 (b) 1.012 (c) 18.5° (d) 20.98°
13. power, $R = 0.004811D^{-2.016}$
    (a) 3.91 ohms (b) 0.193 ohms (c) 0.057"
15. exponential, $P = 14.6516(0.999963)^x$
    (a) 8.64 lb/in^2 (b) 14.65 lb/in^2 (c) 29056 ft.
17. (c) The semi-logarithmic graph gives nearly a straight line for data after 1964, but the data prior to 1964 does not come near the line.
    (d) exponential, $D = 218.443(1.1053^x)$
    (g) 398.3, 1084.2 (in billions of dollars)
    (h) In year 2010: 21863.1 (in billions of dollars)

Exercise 12.3

1. (a) 0.6826 (b) 0.4772 (c) 0.0014
3. Area = 0.6826 Statement: "(1/√(2π))e^(−x²/2)"→Y_1
5. (a) 0.4861, 48.61% (b) 0.9141, 91.41%
   (c) 0.0344, 3.44% (d) 0.9280, 92.80%
7. (a) 81.84% (b) 31.57% (c) 56.75% (d) 24.84%
9. (a) 0.2120 (b) 0.5403 (c) 0.0808 (d) 0.5000
11. (a) 756 (b) 641 (c) 162 (d) 1098
13. (a) 816 (b) 89 (c) 117 (d) 1083
15. (a) 0.68 (b) −1.65 (c) −1.65, 1.65 (d) −2.58, 2.58
17. (a) Between 0.4989 and 0.5071 (b) Between 0.4981 and 0.5079
    (c) Between 0.4966 and 0.5094
19. (a) Between 1.186 and 1.224 (b) Between 1.180 and 1.230
    (c) Between 1.171 and 1.239

Exercise 13.1

	right hand limit	left hand limit	limit
1.	−2	−2	−2
3.	4	4	4
5.	6	6	6
7.	1	1	1
9.	−∞	∞	does not exist
11.	1	1	1

13. 0.750
15. −3.16
17. 2.72
19. 4

Exercise 13.2

	Derivative when x =					
	−2	−1	0	1	2	
1.	2	2	2	2	2	Derivative is a constant 2.
3.	−4	−2	0	2	4	Derivative is twice the x-value.
5.	−32	−4	0	4	32	Derivative is 4 times the cube of the x-value.
7.	−1/4	−1	undef	−1	−1/4	Derivative is negative of reciprocal of x squared.

9. For $x < 1$, derivative < 0, f(x) decreasing; for $x > 1$, derivative >1, f(x) increasing.
11. For $x < -2$, or $x > 2$, derivative < 0, f(x) decreasing;
    for $-2 < x < 2$, derivative > 0 positive, f(x) increasing.
13. For $x < 0$, derivative > 0, f(x) increasing; for $x > 0$, derivative < 0, f(x) decreasing.
15. Horizontal tangent line at x = 2.5.
17. Horizontal tangent line at x = 0.0

19.

s =	−10	−2.5	5.0	17.5	30.0	s is linear, increasing
v =	5.0	5.0	5.0	5.0	5.0	v is constant and positive.

21.

s =	60	90	90.25	90.24	90	s is increasing, then decreasing
v =	44	4.0	0.0	−0.8	−4.0	v is positive, then negative

23.

x =	35	45	50	55	65	The cost decreases for $x < 50$ and increases for $x > 50$.
der =	−30	−10	0	10	30	Minimum cost of $2000 occurs when x = 50.

25. 440, 450, 460, 470

Exercise 13.3

1. $y = 3x + 0.25$
3. $y = -1.14x + 2.81$
5. $y = -0.67x + 10.82$
7. $y = 109.2x - 331.1$
9. derivative > 0 for $x < 4$ and f(x) increasing; derivative <0 for $x > 4$, f(x) decreasing.

11. derivative > 0 for $-1/2 < x < 0$, $x > 0$, f(x) increasing
derivative < 0 for $x < -1/2$, $0 < x < -1/2$, f(x) decreasing

13.-17. (a) When derivative > 0, function is increasing; (b) derivative < 0, function decreasing; (c) At x-intercepts of derivative tangent line to graph is horizontal.

13. Graph of derivative is straight line.

17. At $x = 5$ and $x = -5$, graph of derivative becomes undefined.

19. 200, ball reaches maximum height, derivative > 0 for $x < 200$, ball is rising.

21. second derivative < 0 for $x < 0$; graph concave down
second derivative > 0 for $x > 0$; graph concave up

23. second derivative > 0 for $x < -1.63$, $x > 1.63$; graph concave up
second derivative < 0 for $-1.63 < x < 1.63$; graph concave down

25. −3.741657, 3.741657

27. −2.504978, 2.504978

Exercise 14.1

1. $y = \frac{3}{2}x^2 - 7x + 4$

3. $y = \frac{1}{3}x^3 + x^2 - 3$

5. $y = \frac{1}{4}x^4 + 3$

7. $y = \frac{2}{x} + 6$

9. $y = (x^2 - 2)^4 - 2$

11. $y = x^3 + x^2 - 5$

13. $y = (3x + 5)^8 + 14$

15. $i = 2t^2 - 0.2t^3 + 2.5$

Exercise 14.2

1. 24, 28, 30; lower bound

3. 23.6, 21.1, 19.6, 18.8; upper bound

5. 5.1, 11.4, 15.5, 17.8; lower bound

7. 3.67, 2.11, 1.42, 1.12, upper bound

9. Make the following changes in program ARECTG1:
   - Change line 8 from $A \rightarrow X$ to $A + H \rightarrow X$
   - Change line10 to LINE(X–H, 0, X–H, Y_1)
   - Change line 11 to LINE(X, 0, X, Y_1)
   - Change line 12 to LINE(X–H, Y_1, X, Y_1)
   - Change line 15 from If X<B to If X<B+H

13. 40, 36, 34; $30 < \text{area} < 34$ (exact area = 32)

15. 10.1, 14.3, 16.2, 17.1; $17.1 < \text{area} < 18.8$ (exact area = 18)

17. 45.6, 31.6, 25.6, 22.9; $17.8 < \text{area} < 22.9$ (exact area = 20.25)

19. 0.23, 0.39, 0.56, 0.69; $0.69 < \text{area} < 1.12$ (exact area = 0.875)

21. 21.36, 23.64, 25.52, 25.76; exact value = 26

23. 40.96, 51.84, 61.47, 62.73; exact value = 64

25. −44.6, −42.1, −40.6

27. (a) 84.48, 85.30 (b) 25.92, 5.64 (c) Area from 0 to 12 is algebraic sum of positive area from 0 to 8 and negative area from 8 to 12 (within limits of accuracy).
29. 0.835, 0.919, 0.984, 1.008; exact area = 1.00

Exercise 14.3

1. 32, 32, exact = 32, error = 0.0%, 0.0%
3. 17.7188, 17.9824, exact = 18, error = 1.56%, 0.10%
5. 21.5156, 20.3291, exact = 20.25, error= 6.25%, 3.91%
7. 0.98865, 0.90575, exact = 0.875, error = 12.99%, 3.51%
9. 17.75, 17.1875, 17.04688, 17.01172; errors = 4.41%, 1.10%, 0.276%, 0.0689%; doubling N decreases error by factor of 4.
11. 0.0, 3.99, −3.99; $\int_{-2}^{2}(x^3-4x)\,dx = \int_{-2}^{0}(x^3-4x)\,dx + \int_{0}^{2}(x^3-4x)\,dx$
13. 120.12, −120.12; $\int_{1}^{3}6x^3\,dx = -\int_{3}^{1}6x^3\,dx$
15. PROGRAM: SIMPRULE

```
0 → S
1 → K
Prompt A
Prompt B
Prompt N
(B−A)/N → H
A → X
S + Y1 → S
B → X
S + Y1 → S
A + K*H → X
S + 4Y1 → S
K + 1 → K
Lbl A
A + K*H → X
S + 2Y1 → S
K + 1 → K
A + K*H → X
S + 4Y1 → S
K+ 1 → K
If K ≤ (N−1)
```

```
Goto A
S*H/3 → S
Disp "Value is"
Disp S
```

17. 32, 32; no error
19. 18, 18; no error
21. 20.25, 20.25; no error
23. 0.9011, 0.8781; error = 2.98%, .35%
25. 66937 ft-lb
27. F = 0.1767k

Exercise 14.4

1. 5.436
3. 0.2195
5. 0.9701
7. 39.27
9. 15.63
11. 18.7
13. 5.33
15. 13.7
17. 0.284
19. 17.7
21. 1067 ft^2
23. 97.8 km

Index